Guide for Commissioning Building Electrical Systems

By: Mike Starr, PE

Disclaimer and Information

This guide is provided for personal use only. No part of this document may be reproduced or transmitted in any form or by any means, electronic or mechanical.

This guide is for informational purposes only. Although precaution was taken in preparation, the Author assumes no responsibility for errors or omissions, nor does he offer any expressed or implied warranty about the accuracy of information. The Author does not accept liability for loss or damage caused by using the information contained herein.

If you are reading this guide in an electronic format, reference content links are hyperlinked for your convenience. If you are reading this guide in a physical form, web addresses will need to be manually typed into your web browser. The hyperlinks are to websites controlled and available by outside sources; the Author has no control of the content. These web address references were accessible at the time of publishing but may not be available should the website owners decide. This guide makes no guarantee regarding the availability or accuracy of the information contained on those websites – they are supplemental, for reference only.

The names of the trademarked/copyrighted software and products in this guide are for editorial purposes only and benefit the respective trademark/copyright owners. The terms used in this guide do not intend to infringe on the trademarks and copyrights. All product and company names mentioned in this guide are trademarks (™) or registered trademarks (®) of their respective holders. The use of them does not imply any affiliation with or endorsement by them. All company, product, and service names used in this guide are for identification purposes only. Likewise, all hyperlinks to web addresses are of similar intent as trademark/copyright – for reference only and without association to those companies. This guide is an independent publication.

R1.1

To my brother and best friend, Mark Laufer.
Thanks for showing me the way in life.

Table of Contents

Summary of Figures

Summary of Tables

Preface

Welcome to your *Guide for Commissioning Building Electrical Systems*!

From the Author

My name is Mike Starr. I am an Electrical Engineer, and I started working in the electrical industry in the year 2010.

My career focus has been designing electrical systems for buildings and campus environments. I have also been involved in the commissioning of electrical systems. However, instead of taking an official commissioning course program, the industry trained me. By that, I mean:

I have been behind the desk performing reviews of plans and specifications, writing checklists and functional test scripts, and coordinating load bank plans. I have spent time on job sites with experienced mentors training me on how to perform detailed equipment inspections, configure power quality meters, and operate critical equipment. I have also grown to be a mentor, training other Engineers to design and commission electrical systems. The commissioning projects that I have been a part of amount to extended hours on remote job sites, inspecting, witness testing, and re-testing electrical systems.

So, it is through my project experience, up to this point in my career, that I would like to walk you through the process and provide my suggested implementation for commissioning of building electrical systems.

Visit www.nebulous-llc.com for contact information.

Target Audience

This guide is recommended for:

1. College Students: whether you are an Electrical Engineering, Architectural Engineering, or a Construction Administration Student
2. Electrical Designers and Engineers, Architects, Project Managers, Commissioning Agents, Contractors, and even Electrical Maintenance Staff

Really, anyone in the construction industry with an interest in electrical systems will likely take something away from this guide.

Mike Starr, PE

Here are two considerations to help gauge the assumed experience level of the reader:

1. Since we have an electrical subject emphasis, a basic understanding of electrical building systems is helpful. For example, at a high-level, you can navigate an electrical one-line diagram. Ideally, you are also familiar with reading electrical construction documents. Not that we will be reviewing electrical plans to any extent in this guide, but this knowledge level makes the guide most useful.

2. On the other hand, your electrical knowledge does not need to be so in-depth that you know the entire electrical code; nor do you need to be an expert interpreter of electrical wiring diagrams. However, understanding electrical equipment enough to distinguish panelboards from Uninterruptible Power Supply (UPS) systems or generators is considered base knowledge when using this guide.

Purpose

This guide hopes to offer you a unique perspective on electrical commissioning, with the following goals:

1. Help develop more industry expertise for those Commissioning Agents already testing electrical systems
2. Provide insights for building Owners and project partners to enhance industry coordination for electrical building systems
3. Inspire Electrical Design Staff and other individuals in the electrical industry to consider commissioning as part of their career path
4. Generate Project Team conversations about the electrical commissioning process

Scope

With a mentor-based approach, this text aims to help you understand the commissioning process and provide recommendations for successful projects. This guide offers insights applicable to general building types, such as commercial and office buildings, and advanced building types, such as healthcare facilities and data centers. We will not cover commissioning for specialty applications such as food plant processes, power plants, microgrids, SCADA/DSCADA, Distributed Control Systems (DCSs), utility substations, etc. However, the content could still be used as a template and adapted by the reader to fit similar unique applications.

This guide may apply to many electrical installations, but our examples focus on 60Hz, 3-Phase, Alternating Current (AC) power systems (less than 1000V). Many of the insights are useful for commissioning systems up to 35kV (non-utility installations).

Before narrowing the discussion to electrical systems, we outline the benefits of commissioning, examine factors for project schedules/budgets, and detail the phases of a typical commissioning scope (from an electrical point of view). Following a reminder about the importance of electrical and job site safety, instead of general guidance, we discuss specific electrical systems and the appropriate considerations to take for inspecting and testing them. We also overview power characteristics, test equipment, and highlight valuable tips for checklists and test scripts.

Throughout the chapters, references are suggested for further information and confirmations. It would be redundant to over-document recommendations already thoroughly described in outside references, such as NETA and NECA, so they are simply listed throughout. For example, topics such as acceptance testing procedures.

The National Electrical Code (NEC, NFPA 70) and NFPA's other codes/standards are available free directly online from NFPA's website – the user just needs to make a free login.

NFPA Free Access: https://www.nfpa.org/Codes-and-Standards/All-Codes-and-Standards/Free-access

To understand which codes apply for the jurisdiction, the online application CodeFinder (also by NFPA) is another free helpful reference.

NFPA CodeFinder: https://codefinder.nfpa.org/

For codes and standards other than NFPA, you may need to borrow or purchase documents from outside sources to investigate them. Having these resources available for review as you read is helpful, but otherwise, they are purely supplemental.

Industry Verbiage

Depending on which commissioning standard is being referenced, the exact term for the Commissioning Team varies: Engineers, Agents, Group, Provider, Authority, etc. For our purposes, the reference to these team members means the same thing: the industry professionals who commission building systems. See Annex B for an alphabetical list of acronyms and abbreviations used throughout.

Mike Starr, PE

One Engineer's Point of View

Of course, this guide's content is based on the Author's experiences and developed industry knowledge. His expertise and guidance will not align with every project type. Ask any Engineer what the right solution is, and their answer will depend on the circumstance and their exposure to various project types. Please do not proceed with any aspect of commissioning just because this guide makes a suggestion. Do open the cited references and weigh all of the industry guidance available, along with your own experiences. Performing due diligence for the commissioning scope of work is the Commissioning Professional's responsibility.

As with any reference, this guide seeks to provide complete information, but it is certainly only a portion of the knowledge and experience needed to properly commission building electrical systems. Examples used throughout the guide will not cover every condition, but hopefully, they demonstrate the types of considerations a Commissioning Agent uses when making decisions or recommendations.

This guide uses specific Manufacturers in examples only because the Author is familiar with them. Other Manufacturer options could be specified for equally successful projects. This guide does not seek to promote any single Manufacturer over another. When multiple product examples are listed, they are in no particular order.

The solution to a commissioning problem typically originates at the design stage. For this reason, in addition to the Commissioning Agent's role in the design phase activities, the Author's viewpoint has a balance of design considerations embedded in them.

Thank You

First, a sincere thank you to my personal and professional life mentors. This guide would not be made possible without you. In particular, those of you who helped review this guide in part or whole. I am forever grateful for your support!

To the reader: thank you for considering this guide to learn more about commissioning building systems. I hope the content meets your expectations and provides practical guidance for electrical commissioning, not just in theory but for your real-world projects.

Chapter 1: Introduction

What is Electrical Commissioning?

Origin

The building commissioning process came about in the 1980s. It is a method to confirm the performance of engineered systems. In particular, Commissioning (Cx) is a process rather than any single activity. For example, if a building includes a back-up generator, the commissioning process would confirm the installation and all operating characteristics, including failure modes. If the generator system does not go through commissioning, the Installing Contractor and Manufacturer Teams usually only complete routine confirmations. Without commissioning, it is unlikely for rigorous functional testing or quality check documentation beyond general start-up activities.

Initially, due to code-required functional testing, the commissioning process primarily focused on:

- Mechanical systems: heating, ventilation, air conditioning, and piping systems
- The building enclosure (also referred to as the building envelope)

In many jurisdictions, electrical commissioning is voluntary and has only started to gain traction beginning in the early 2000s – this may be because mechanical systems and building envelopes significantly impact energy savings and occupant comfort. Operational savings and happy employees are reason enough for Owners to, understandably, pay close attention to the building mechanical and enclosure systems. Since lighting systems also relate to energy and occupant comfort, lighting has had some voluntary commissioning since the 1990s. Starting in ASHRAE 90.1-2010, lighting systems have functional testing requirements.

Energy usage and building system performance are key factors driving the commissioning process, which may be applied to new and existing buildings. An existing building may even need to be re-commissioned, which means it has been commissioned in the past. If an existing building never went through commissioning at the time of original construction, the industry refers to this as a retro-commissioning project. Commissioning of existing buildings may provide energy savings or correct deficiencies in the building control systems.

When looking to understand the energy codes and other requirements that are enforced, COMCheck is a helpful free software tool for confirming the enacted building energy requirements: https://www.energycodes.gov/comcheck

Mike Starr, PE

In many jurisdictions, a completed COMCheck analysis is a Design Team requirement for the building permitting process.

Beyond energy compliance for mechanical systems and the building enclosure, many Owners are electing to include electrical commissioning services for their buildings – the sections in this chapter overview why this is becoming more common.

Entire Building Commissioning

The various commissioning organizations term it differently: whole building, total, full-service, or complete building commissioning. Entire building commissioning is a risk management strategy. Choosing to proceed with commissioning all building systems comes with energy and maintenance savings opportunities. What follows commonly translates to lower repair costs, longer equipment life, happier building occupants, and less downtime. Commissioning helps identify issues before the Installing Contractor's warranty expires, which is typically twelve-months from the initial building occupancy or substantial completion, no matter if the building goes through commissioning or not. Today, many building Owners choose a full-service commissioning approach, which encompasses electrical systems. Some companies require a capital construction project for their organization to employ the entire building commissioning process.

QA/QC

Commissioning is easy to confuse with another construction process: electrical acceptance testing. The Owner's insurance company may have specific acceptance testing requirements listed in their standards. For example, it makes sense to perform insulation resistance testing before energizing feeders. If the project does not include formal acceptance testing, then the Installing Contractor is only required to complete minor checks before energizing equipment. Typically, the minor checks are safety-related, not system performance-driven.

Together, the commissioning process and acceptance testing are a means of Quality Assurance and Quality Control; or shorthand as QA/QC. The commissioning process assures that the systems operate as designed, and acceptance testing is the quality control mechanism.

Typical Electrical Systems to Commission

Now that building Owners are finding value in commissioning electrical systems, the industry has been trying to shape which systems to test and what level of rigor each system should receive. Critical electrical systems are given the highest priority. Critical can mean life-safety and/or systems that drive business continuity (mission-critical). For example, powering a production line in a factory that losses the entire product if power is interrupted or takes a significant amount of time to restart the process. If the project budget is sufficient or a risk assessment determines necessary, it may make sense to commission non-critical systems as well. Reference NFPA 70E, Annex F: Risk Assessments and Risk Control. Annexes in NFPA 1600 may also help assess business risk.

A common approach to determining which electrical systems to commission is using an electrical one-line diagram to trace the power pathways serving critical circuits. The Owner should work with the Design Team to determine which systems are most appropriate for commissioning. The industry enjoys grouping systems: normal power, emergency power, UPS power, etc. However, to better delineate the project scope, breaking down systems into equipment types is recommended.

These are some of the electrical systems typically considered:

- Generators
- Transformers
- Power Distribution Units (PDUs)
- Emergency Power Off (EPO) systems
- Uninterruptible Power Supplies (UPSs)
- Metering or Power Monitoring Systems (PMS)
- Power transfer equipment (automatic or manual)
- Low-voltage systems (security, data, gas detection, etc.)
- Life safety systems (egress lighting, fire pump, fire alarm, etc.)
- Lighting control and spot checking of general-purpose receptacles
- General power distribution equipment (pad mount equipment, switchgear, switchboards, panelboards, etc.)

Motor controls (VFDs, combination starters, etc.) are typically commissioned, but in the Mechanical Commissioning Agent's scope of work since they interface with these motors and controls significantly during their system testing.

The Design Engineer determines which specific systems are commissioned (QA) and/or acceptance tested (QC). In order to avoid higher insurance premiums, the Owner's insurance carrier may have acceptance testing requirements for electrical equipment. The electrical acceptance testing activities (QC) may test all systems, such as motors, batteries, surge protection, control power transformers, grounding systems, fiber-optic cables, etc.

Mike Starr, PE

For further information on electrical commissioning, consider reviewing:

- ASHRAE 0: The Commissioning Process
- BCA Building Commissioning Handbook
- NECA 90: Commissioning Building Electrical Systems
- ASHRAE 202: The Commissioning Process for Building Systems
- NFPA 3: Standard for Commissioning of Fire Protection and Life Safety Systems
- NFPA 70 (NEC) Annex F: Availability and Reliability for Critical Operations Power Systems; and Development and Implementation of Functional Performance Tests (FPTs) for Critical Operations Power Systems
- NFPA 70B, Chapter 31: Electrical Preventative Maintenance (EPM) from Commissioning (Acceptance Testing) Through Maintenance
- NFPA 72: National Fire Alarm and Signaling Code

When Should Electrical Commissioning be Considered?

The team must consider if commissioning is required. Even if it is not required, there may be a benefit to the project. For instance, commissioning likely results in energy savings, possible tax credits, or may qualify for local utilities' energy incentive programs. Additionally, commissioned building systems typically benefit from lower maintenance issues (see the Entire Building Commissioning section).

In some cases, due to the project budget, a standard or enhanced Contractor quality program may meet the project's needs. Examples of projects that may not warrant the additional expense of commissioning include temporary installations or projects with short construction schedules. In those instances, commissioning may be an excessive expense or constraint on the project schedule. Depending on the level of comfort and any pre-established Preventative Maintenance (PM) contracts, the Owner's Maintenance Staff may find it acceptable to avoid the additional time and cost of commissioning.

The following sections review possible reasons to include electrical in the Commissioning Provider's scope.

<u>Standards Driven</u>

Project sustainability targets is a common reason to include commissioning services. For example, Leadership in Energy and Environmental Design (LEED). Other standards are growing in popularity and may also have commissioning requirements, such as the WELL Building and Passivhaus standards. Commissioning may be needed to complete a program requirement or as an enhanced follow-through measure to ensure design compliance during the construction phase. An Owner might also consider electrical commissioning when significant building code changes have been recently adopted. A Commissioning Team helps the Design Team confirm the project meets all of the current requirements.

Above all, due to the high cost of downtime, electrical commissioning is very common for data centers. These applications seek to align with standards such as The Up-Time Institute, ANSI/TIA 942, BICSI, NFPA 75, and ASHRAE 90.4. In these mission-critical applications, energy efficiency is a priority. It is also unacceptable to have a loss of power to the building's critical operations.

Confirming Equipment Used for Service Continuity

The number one reason why institutions and organizations consider electrical commissioning, aside from code or standard requirements, is to confirm proper operation of the systems they have purchased to avoid downtime (loss of power). If you have ever been working in an office building when the power goes out in the middle of the day, you know what happens, there is a lot of standing around. If there is no building power, that likely means there is no network connectivity or building cooling. All productivity pauses or slows to a crawl until power returns. Depending on the business type, the cost of downtime may be high. In healthcare facilities, if the power does not return quickly, human life could be at risk.

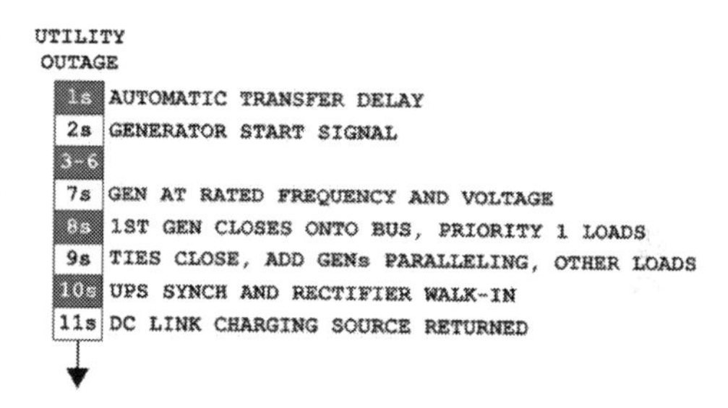

Figure 1: Example Source Outage Timing

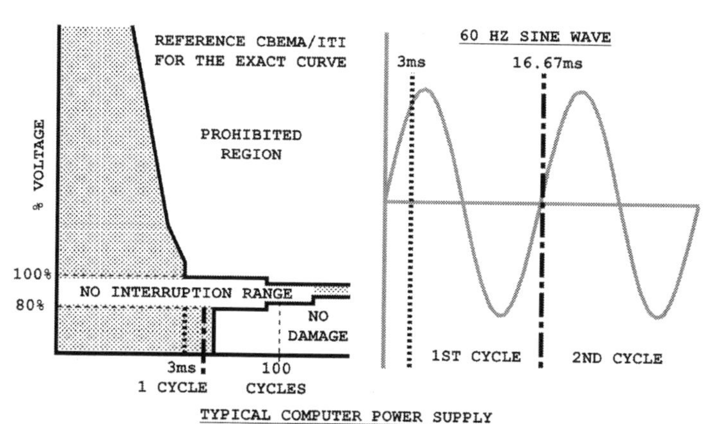

Figure 2: CBEMA/ITI Curve and 60Hz Sine Wave

As shown in Figure 2, it does not take very long before for a typical computer power supply is at risk of powering off due to loss of input power. Many other digital electronics perform similarly and are at risk with even short-duration power anomalies (less than one cycle).

Even ordinary commercial customers, who are not mission-critical, have business-critical functions that benefit from maximum availability to their users. Ultimately, the goal is common for mission and business-critical operations: limit power outages to scheduled outages only, because no one wants an expensive unplanned outage.

When considering electrical service continuity, the Eaton Blackout Tracker overviews causes of power outages within a given state or region and is a handy tool for clients and Project Team members to review and discuss:

https://switchon.eaton.com/plug/blackout-tracker

If electrical system malfunctions are low impact to business continuity and building certifications are not a requirement, commissioning may not be necessary.

Non-Standard Equipment

Outside of the mission-critical and healthcare market sectors, the most compelling reason to commission systems is when the design or equipment is non-standard. Some examples include:

- New installations that have modern microprocessor-based controllers and relay protection
- Critical equipment that depends on custom Programmable Logic Controllers (PLCs) or relays protocols with Human Machine Interfaces (HMIs) for automatic source transfer schemes
- Life-safety applications with emergency back-up systems that must return power in under 10 seconds
- When there is a function change, such as higher-density electrical loading, in a laboratory, computer room, or similar unique application

Redundancy increases fault or outage tolerance and helps to "maintain business as usual" when building maintenance needs to occur. For example, a building may have a Main-Tie-Main (MTM) service entrance switchboard, where two power sources are tied together (isolated by the open tie breaker) to serve a common group of loads. A building with this type of equipment likely has independent power sources serving each main breaker. On the surface, more redundancy sounds like a great idea. Actually, the increased resilience results in greater system complexity for Building Operators and Maintenance Staff. This hurdle should not deter designs from using redundant systems to improve reliability, but it should highlight the importance of confirming those systems via a commissioning process before building occupancy.

System Complexity

Other example applications that may be standard, but due to complexity likely benefit from a commissioning process:

- Redundant equipment and system configurations
- Specialty equipment, such as metal-clad switchgear
- Newer technology, such as lithium-ion battery storage
- Projects addressing existing deficiencies or scheduled replacements
- Multi-building campus environments that require consistency for maintenance sake
- High-altitude project sites, or where there are other environmental stresses on the electrical equipment
- Transfer equipment: Due to the critical nature of transfer equipment, even non-custom equipment is usually recommended to be commissioned, along with the up/downstream systems connected.

Project Scale

Construction Managers (CMs) will sometimes seek third-party due diligence via a commissioning process. For very large, multi-year projects, a dedicated Commissioning Provider might be used for the design phase. They are responsible for writing the commissioning documentation. Then, this representative oversees an independent Commissioning Agent performing third-party commissioning services in the construction phase.

Baselining Systems

The commissioning process seeks to establish baseline performance criteria. The Owner's Maintenance Staff then has the data to reference for proper maintenance over the life of the systems. For example, a UPS system may not see peak loading for many years. The commissioning process captures the UPS peak loading baseline data on Day 1 to provide comparisons as the real building load grows overtime. Likewise, the UPS battery runtime may also be captured to understand the maximum back-up time. The data provided from the testing process is reviewed during maintenance programs to monitor equipment health.

Having baseline information is a valuable troubleshooting tool. Commissioning data specific to a facility's equipment and the interconnected system is even more valuable than individual Manufacturer's equipment data. In fact, if the building has any system issues over the lifetime, one of the first questions for the Owner is, "Do you have building commissioning data?" For example, if a system anomaly shows up during commissioning, the documentation may describe the remedy.

Projects that choose to include commissioning may also find marketing advantages for having this baseline data. For instance, co-location data centers renting out Information Technology (IT) cabinet space may have clients who require commissioned computer racks. That potential data center tenant may request a copy of the final commissioning report before signing a contract.

<u>Owner Training</u>

It is encouraged for the Owner's Maintenance Staff to walk the project site as much as possible throughout the construction process. Further, it is usually helpful to the Maintenance Staff if they participate in the functional testing of the systems.

It makes sense that building systems will operate more efficiently by training the operators. There may need to be multiple training sessions for the various Maintenance Team members' who work on different shifts throughout a 24-hour day. Although some form of training is typically included by the construction process already, formal Owner training initiated by the commissioning process is a significant benefit of bringing electrical systems into the commissioning scope.

As buildings and systems are new company assets, many Owners find construction to be an excellent opportunity to review their in-house maintenance program and make process improvements. The Owner Team walks away from the construction process with useful documentation and possibly even training videos for existing and future staff.

A more detailed discussion on Owner training is included in the Post Construction section.

Chapter 2: Project Management

Scope

Team Structure

The most common arrangement for contracting commissioning services is direct with the Owner. This approach helps keep Owner Representatives up-to-speed when the project encounters challenges. Optionally, if the Owner is comfortable with the working relationship, the Commissioning Team could also be contracted through a Construction Manager (CM). No matter the contract arrangement, maintaining objectivity is paramount to success. Therefore, as required by most commissioning standards, the Commissioning Provider should report directly to the Owner.

If the project is pursuing the LEED Fundamentals Credit and the building is under twenty-thousand square feet, the Commissioning Provider may be from the Design or Construction Team. However, the LEED Enhanced Credit requires the Commissioning Provider to be a third-party, separate from the Design and Construction Team companies. The US Green Building Council website clearly outlines the LEED requirements for commissioning. LEED requirements vary over time, so be sure to review the most current version of the standard – including the exact contract arrangements that are required. Likewise, for other building standard commissioning requirements – confirm these directly with the credential authority.

Even if the project does not seek official certification, having the Owner and Project Team commit to a sustainability standard pushes forward essential "if we have time" tasks that get lost in the pace of the construction schedule. For example, the Project Team's attention and input on the Owner's comprehensive Preventative Maintenance (PM) plan. All of these expectations should be outlined in the Owner Project Requirements (OPR) and/or Basis or Design (BOD).

Unless restricted by a commissioning standard, if the Design or Contractor Teams have relevant commissioning experience, there may be an opportunity for cost savings by not having to introduce a separate Commissioning Team to the project. If the Commissioning Team member is from the Design or Construction Team, ideally, that new team member has not yet worked on the current project scope. For example, if the Design Team's company commissions the project, there is both a Design Engineer and a separate qualified Commissioning Agent; this way, the project still benefits from a fresh perspective. The commissioning scope is a non-standard value-added service for design and construction, so the Owner would need to request a proposal for this additional work from the team if desired.

Mike Starr, PE

Credentials

Commissioning Team members might have expertise in all building systems (controls, mechanical, electrical, etc.). Successful Commissioning Providers also bring knowledgeable staff who may have experience with multiple system configurations, which adds to and helps diversify the Design and Contractor Team knowledge base. For larger and more complex building types, it is helpful if the Commissioning Team has members that specialize in electrical systems. Ideally, the Electrical Commissioning Agent will have familiarity with testing and maintenance of electrical systems, but also a strong knowledge of current building codes and industry standards.

It is a good idea for Owners to go through an interview process with prospective Commissioning Providers. Likewise, when determining an existing space's overall scope, it is prudent to do a walk-through of the project. When evaluating companies to perform commissioning, several industry organizations offer commissioning credentials to individuals and entire companies. To name a few:

- Certified Commissioning Provider (CCP)
- Certified Commissioning Authority (CxA)
- Building Commissioning Professional (BCxP)
- Certified Building Commissioning Firm (CBCF)
- Qualified Commissioning Process Provider (QCxP)
- Certified Building Commissioning Program (CBCP)
- Commissioning Process Management Professional (CPMP)

Some of the commissioning programs offer generalist and specialist versions of their certifications. Many of the programs require a minimum number of years working in a commissioning role. The required minimum number of years' experience is less for those who have a professional license and supporting education or experience. The industry recommends that the Commissioning Team have at least one Professional Engineer (PE) team member. The licensed professional then takes on the responsibility of reviewing code compliance and acceptance testing data for accuracy. This individual could be one or more of the commissioning reviewers (from the design phase). These types of staff qualifications should be considered differentiators in the Owner's Commissioning Team selection process. For more information, reference ASHRAE 0, Annex D: Commissioning Process Request for Qualifications. To identify the best team for a given project, Owners sometimes select the Commissioning Team by qualifications first. Then the scope/fee is refined to meet the project's commissioning needs.

Commissioning Role

The commissioning role takes on responsibility for verifying the installed systems and whether those systems operate as designed by the Engineer of Record (EOR). They typically take a hands-off approach, working alongside the systems' Installing Contractors via in-person inspections using checklists (Contractor completed) and test scripts. The Commissioning Agent should avoid directing project changes. Avoid advising Contractors to make changes, even if the planned install has deficiencies; this is not just a best practice, but contractually the Commissioning Agent also does not have the authority to change the design. The proper protocol is for the Commissioning Agent to provide suggestions for change/correction or simply identify deficiencies without recommendations, and the Design EOR will decide the path forward with the Owner.

For example, rarely does the EOR publish their required settings for transfer equipment (time delays, thresholds for voltage and frequency drop-out/pick-up, etc.). Default settings may be acceptable for basic electrical system designs, but otherwise, custom settings may be needed; especially, when transfer equipment feeds other downstream transfer equipment. The Commissioning Agent may have recommendations on settings that will work for the installation, but ultimately it is up to the Design Team to determine the final settings. Including proposed settings or the transfer equipment's available default settings in a Request for Information (RFI) to the EOR may result in a faster reply. The Design Team might consider transfer inhibit signal wiring between series wired transfer devices to avoid a settings conflict.

Commissioning and Acceptance Testing

There are two contractual approaches for electrical system QA/QC:

- Single contract performing acceptance testing and commissioning
- Separate contracts performing acceptance testing and commissioning

The single contract approach requires the Commissioning Provider to have technicians, often NETA or NICET certified, performing both acceptance testing and commissioning services. The combined arrangement (single contract) is less common, but can be a very beneficial. Since they spend their days performing hands-on testing of electrical systems, the individuals working in these companies are highly skilled. Especially for occupied installations, which normally have short power outage opportunities for performing the work, the combined acceptance-commissioning relationship may make construction scheduling easier. A mixed Acceptance-Commissioning Provider contract is with a single company, but may also be an acceptance testing company sub-contracting a Commissioning Team or vice versa.

More commonly, the Electrical Contractor is responsible for the acceptance testing scope and hires an independent acceptance testing service group, or performs acceptance testing themselves if the design specifications allow them to self-perform the acceptance testing scope. Traditionally, this is the case because the acceptance testing requirements are usually included in the design specifications; then, a hands-off (witness only) Commissioning Provider is hired separately by the Owner. Commonly, Mechanical Electrical Plumbing (MEP) design firms include a witness only Commissioning Team. A witness only (consultant) Commissioning Team is still very knowledgeable and generates all of the same commissioning documentation. However, instead of physically opening and closing breakers themselves, they direct Installing Contractors and Vendors for equipment operation and setting verifications.

Commissioning Plan

Once contracted to perform work, the Commissioning Team assembles a document called the Commissioning Plan, which is a roadmap to executing the commissioning scope of work. Suggested topics to cover in this document include:

- Describe Owner training
- Outline the planned commissioning process
- Clarify if official project certifications are being pursued
- Establish preliminary commissioning meeting dates and times
- Provide a high-level draft of the project schedule with major milestones
- Include example scripts for major systems to demonstrate the level of rigor
- Define commissioning roles and responsibilities
  - For example: Owner, Electrical Contractor, Controls Contractor, etc.
- Outline general building info
  - For example: number of floors, type of construction, occupancy, etc. This type of information should be readily available from the Design Team.
- List the team members from each company, including contact information (email/phone) and team member roles in the commissioning process
- Establish equipment types and systems to be tested
  - If some systems are to be sample tested, in place of 100% testing, identify the percentage sampling rate. For example, verifying addressable light fixtures may be a sampling activity rather than 100% confirmations.

Example generic plan (PDF download):

https://www.cacx.org/resources/cxtools/documents/05_CommPlanTemplate_EDR.pdf

To assist with scope delineation between various entities, some Project Teams find it helpful to establish a responsibility matrix and include it in the Commissioning Plan. This matrix lists all equipment/systems to be commissioned and identifies who is responsible for project construction and commissioning activities. As the project progresses, the responsibility matrix may require updates.

ASHRAE 0 and 202 Annexes have samples of the many document types used on commissioning projects, including suggestions on sampling rates.

Schedule

Pull Planning

In the early stages of construction, the overall build schedule is slowly informed by equipment and material lead times (ordering, submittals, shipping, etc.). Still, it is essential to start shaping the commissioning schedule, which the Commissioning Provider usually manages separate from, but coordinated with, the construction schedule.

One method to capture the finer schedule details and enhance coordination is a pull planning session (construction planning technique). For this method, the primary contract holder prints a large-format timeline/schedule to hang on the wall. Be sure to identify holidays and Owner black-out days that do not allow project work. Then each Project Team member utilizes sticky notes, categorized by color, to mark significant milestones that need coordination with other team members. Each team member then places these activities in the timeline based on when they believe they need that activity/information to be completed. All team members discuss the resultant schedule and collaboratively revise the timing of activities.

Below is a non-inclusive, electrically focused, list of key commissioning landmarks that may be helpful to identify on the project timeline:

- Equipment start-up dates
- Preliminary Owner training dates
- Factory Witness Test (FWT) dates
- Acceptance testing of major equipment
- Test equipment shipment and pick-up dates

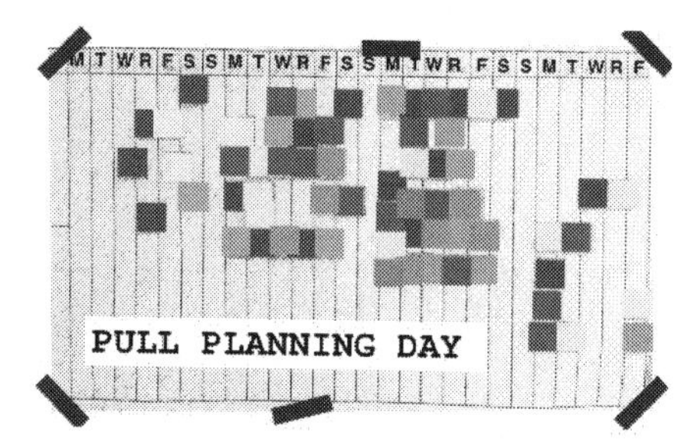

Figure 3: Pull Planning Layout

- The preliminary date that the Test Equipment Plan will be available (more discussion in the Test Equipment Plan chapter)
- Power system study complete, as the settings are critical to safety and the ability to perform some acceptance testing
  - For example, acceptance testing reports require "as-found" and "as-left" settings.
- Building network availability date to confirm electrical monitoring points
- Air conditioning availability date for UPSs, inverter, or station power battery systems
- Fuel tank pressure testing and fuel delivery date for generator(s)
- Functional Performance Testing (FPT) timeline (typically at least one day for each major piece of equipment)
  - For example, a basic transfer switch may only take one or two hours, but a Main-Tie-Tie-Main Secondary Unit Substation (USS) with generator roll-up options may take one or two-days for functional testing.
- Integrated System Test (IST) timeline – typically five or more days. Another possible metric is one day per 1MW of power to be commissioned.
- Owner Team planned activities
  - For instance, if they are waiting for commissioning to be completed to start the install of computers into IT racks for their start-up.

Once all team members have contributed to the pull planning activity, the primary contract holder utilizes project software to make an electronic schedule. This schedule is usually in the form of a Gantt chart, which is a means to visualize tasks and relationships of tasks scheduled over time.

Left-Shift Opportunities

"Left-shift" is a term used in the construction industry. The idea is to look for activities to shift left on the construction timeline, such that they take place earlier in the schedule. The industry is taking steps to left-shift construction and commissioning processes. For example, a project may elect to accept Manufacturer or Integrator factory testing of a product; that would be in place of on-site testing or possibly less on-site testing. Depending on the rate of failure found in the duplicate field testing, left-shift may or may not be helpful.

Shifting activities left helps uncover and resolve potential issues faster. Often these left-shift activities enhance project quality. For example, the Virtual Design and Construction (VDC) process uses Building Information Models (BIMs) to improve coordination in a computer model before physical installation.

Another common industry trend is prefabrication (prefab), which some may consider to be a left-shift opportunity. Contractors use prefab processes to perform work in a quality-controlled environment, such as the Contractor's off-site workshop. The prefab process allows Contractors to stage/modularize the install, all while the building is having steel erected and concrete poured. Then the field teams can install at a faster pace when they arrive on-site. Contractors are prefabricating entire electrical rooms on metal skids, standing-up pre-assembled electrical room walls, designing conduit racks with standard offsets, and creating combined system racks for pipes/ducts/conduits/etc. In some instances, Owners include System Integrators: multiple pieces of electrical equipment ship to an Integrator factory, and the Integrator is responsible for combining sub-systems before they ship to the final job site. An Integrator role may be utilized for unique combinations of industrial motors, modular building structures, building controls, etc.

Commissioning itself is a left-shift opportunity. Together, the Project Team should evaluate the building construction and system test sequences to see if there are opportunities to shorten the project schedule and maintain the same or higher quality level. For example, some switchgear Manufacturers offer to build control systems as a simulator in a portable box or on a desktop computer. For the added Owner cost, this approach allows testing the primary sequences before ever being on-site with the real equipment. Buying a simulator not only allows thorough testing, but it is also an investment in maintenance and operations as a training and pre-test tool over the life of the building.

Example simulator (PDF download): https://www.russelectric.com/wp-content/uploads/16-168-Training-Simulator-Brochure.pdf

Although left-shift is a desire for faster schedules, it typically comes at a higher initial cost for the Owner, so the Project Team should avoid left-shifting without weighing the project impacts first.

Coordination

In the early stages of commissioning, electrical and mechanical systems do not have high interdependence since acceptance testing is focused on a specific piece of equipment or system to ensure proper operation. As systems begin to come on-line though, the electrical system testing must be staged with testing of other systems. For example, a chiller cannot be tested by a Mechanical Commissioning Agent if its power keeps turning off and on. Or if the chiller restarts too many times over a given period, the multiple restarts may cause the chiller's onboard controller to time-out and cause downtime for mechanical testing.

What if it rains? Does the team continue powering the outdoor load bank? In the event of rain or projected rain, the responsible Contractor (not the Commissioning Team) must work with the Owner to determine if they want to risk testing. If the rain is more substantial than a very light-sprinkle, the Contractor decides to reschedule testing more times than not.

The project should identify regularly scheduled commissioning meetings to help with coordination. Often it is helpful to schedule them just before or after other on-site meetings when all Construction Team leaders can attend. Scheduling equipment inspection dates with the Electrical Contractor is also recommended. Coordinating the inspection effort helps the Commissioning Agent avoid logging issues for portions of the installation that the Contractor has (knowingly) not made progress on yet, as they are balancing other site activities. Ideally, the Contractor should communicate if inspections or tests need to be rescheduled, so the Commissioning Team does not make unnecessary site visits.

Where possible, the Commissioning Team may find it helpful to implement functional testing immediately after equipment start-up, when the Manufacturer Service Representative can be present for commissioning. Specifying commissioning to align with start-up activities saves temporary test equipment costs and hopefully resolves issues faster since the technician is already on-site. It is not uncommon for the Commissioning Team to show up to a job site for functional testing and find that this is the first time the Manufacturer Start-up Technician has seen the required test procedure. Ideally, the Vendor doing start-up has seen the test script before test day.

From a scheduling standpoint, installation and removal of load banks can be a significant effort. Also note that certifications, such as UpTime, may require verification testing after commissioning is complete – extending the need for test equipment by days or weeks. These types of activities must be reviewed in detail to build accurate project schedules.

When everyone understands the efforts of the other team members, projects tend to move along more smoothly. Ultimately, the goal of coordination is to get all affected teams and individuals on the same page. Establishing a coordinated schedule is a large part of that effort. Collectively, the team seeks to build realistic project schedules with some flexibility for overnight shipped replacement parts, re-testing, and similar show-stoppers that may occur.

Forming a Commissioning Budget

Many would shy away from providing budget type information within a guide book. However, for the Project Team to have meaningful conversations about the cost-benefit of commissioning electrical systems, it is helpful to know what the Rough-Order-of-Magnitude (ROM) costs might be. The ranges noted in this guide are meant to be conservative and may differ for your market, location, and project type. The actual costs may be higher or lower than this guide suggests. Do not use budget or cost information from this guide for real project work unless you have tested them against historical project costs for confirmation.

The cost ranges provided in this section are based on the year 2020 (US-Midwest); hence, the costs may need to be escalated 1% to 5% higher, per year, for the year you are reading this guide. The reason for the percentage cost increase is due to things like inflation and material cost fluctuations. The specific costs in this guide are from rental catalogs and Vendor quotes seen over time. Of course, no escalation is needed for percentage metrics, but they should still be scrutinized based on your specific application.

Reference industry cost estimating resources or contact service providers directly for further guidance on budgeting for the commissioning scope. Examples of estimating software: RSMeans, Accubid, WinEst, Mike Holt, and Manufacturer pricing lists. Other software options exist (non-inclusive list) and may even be better suited for estimating depending on the application.

<u>High-Level Metrics</u>

Every project is different, so there is no great rule-of-thumb to assume the cost to commission a building. Experienced Contractors may have refined metrics based on their past experiences and historical data; although, the total cost of commissioning varies widely amongst building types and applications. If you have complex systems, there will be a higher cost for commissioning, extending any calculated pay-back period. Note pay-back calculations should consider the cost for services and also the required test equipment. Commissioning services may be lump-sum or time and material-based fees. One first-cost metric giving somewhere to start considering the cost of commissioning is NFPA 70-Chapter 31. The contained reference cost percentages are from the year 1998. They describe the cost of commissioning to be:

- <u>Entire Building Commissioning:</u>
  - Between 0.5% to 1.5% of the *entire construction* cost
- <u>Electrical Commissioning:</u>
  - Between 1.0% to 1.5% of the *electrical system* cost
- <u>Mechanical Commissioning:</u>
  - Between 1.5% to 2.5% of the *mechanical system* cost

According to the US General Services Administration, the cost is between 0.5% to 2.25% of the entire construction cost for an entire building commissioning approach.

Reference the GSA website page titled *"Establishing Initial Budget"* for further detail and notes on cost-benefit analysis:
https://www.gsa.gov/real-estate/design-construction/commissioning/commissioning-program/building-commissioning-process/planning-stage/establish-initial-budget-for-commissioning

Mike Starr, PE

The actual cost for commissioning depends on the building application and system types. For example, if the Commissioning Team is to perform services in a live data center, the tasks will take more time, and the cost may be much higher. If testing is during irregular working hours, such as over the weekend or late at night, Contractor retail labor rates will also likely be higher.

Cost of Commissioning

Testing all sequences for the equipment and systems takes time. In particular, it takes time to re-test equipment if there are issues. Although Commissioning Providers understand there will be re-testing of some equipment, the team does not know how much re-testing will need to take place. There may be an additional fee required if re-testing amounts to the Commissioning Team spending excessive time on-site or if they have multiple repeat site visits.

From the Commissioning Provider standpoint, fee proposals are commonly built-up according to the assumed number of hours per activity. Retail labor rates for Commissioning Agents may range from $100 to $400/hour. For a simple test script, this may take the Commissioning Team only a couple of hours to write, as they have templates from past projects that may be quickly modified. For a custom piece of switchgear, the team would not have template scripts and must develop them. It is helpful if the Manufacturer has a sample test script. In those instances, with customized equipment, it may take the team five or more days to develop a script and revise based on feedback from reviewers. Even if there is just one test script, there are additional labor hours to consider. For example, there are design documents to review, job sites to travel to, reports to write after testing, cloud-based software subscriptions for project setups to pay for, etc. When they are not behind the desk, the Commissioning Agent's working hours each day are longer than a standard eight-hours, as most projects are remote, requiring travel time. It is not uncommon for test days to be extended twelve-hours plus.

From the Installing Contractor's point-of-view, it is crucial to understand the project scope related to supporting the commissioning effort. It would not be unusual for an Electrician supporting the commissioning process to spend three to four times the amount of time they would normally spend on a Manufacturer's start-up. Some Electrical Contractors are becoming more experienced with the commissioning process, so the time it takes might be less. Although it may be difficult to attain during the bid process, it would be beneficial to understand the Contractor's experience with commissioning before selection. The Contractor's commissioning support staff will be filling out checklists, attending additional meetings, resolving installation discrepancies, operating equipment to place the system in the various modes, etc. The General Contractor (GC) or Construction Manager (CM) may also benefit from having dedicated commissioning staff members. All Contractors should communicate with their Manufacturer Representatives to confirm alignment on the anticipated scope for commissioning services. If contracted, the Manufacturer Service Technician stays on-site to assist with functional performance testing, so they may quickly document and resolve any discrepancies.

The best way to establish a project budget is to request an official proposal from a qualified Commissioning Provider.

Cost of Acceptance Testing

If the Electrical Contractor does not already include this scope due to a design requirement, the Owner may ask the Commissioning Team to define the acceptance testing scope. Higher labor rates are common for acceptance testers from $150 to $500/hour. Keep in mind some activities may require more than one tester. As with commissioning, there is travel time, test equipment to maintain/calibrate/transport, report writing, etc. A helpful reference for estimating this scope (try a different web browser if it does not work properly):

http://industrialtests.com/testimator/

Of course, Contractors and qualified testing agencies are your best source for budget numbers. Note there may be a Contractor percentage mark-up for coordination if they are sub-contracting the acceptance testing agency.

Cost of Renting Test Equipment

As we overview in the Test Equipment Plan chapter, test equipment is a significant scope to be coordinated. The Commissioning Agent should prioritize the Test Equipment Plan to help determine the rental costs. The most common arrangement is having the Electrical Contractor be responsible for these rentals; especially, if the Commissioning Provider is a witness only (consultant) representative. The Contractor would then be the sole party responsible for ordering, receiving, hooking-up, operating, and shipping the test equipment back on-schedule. Writing the observations about the Power Quality Meter (PQM) test data may be done by the Rental Supplier, but this service costs extra and may involve generic reporting. It is recommended the Commissioning Agent be the one responsible for the PQM reporting; given, the Commissioning Agent is the one directing the tests and writing the final commissioning report, so they are the most efficient team member to review and comment on the test data. The Contractor performing Infrared (IR) scanning and similar scope is usually responsible for those summary reports.

In some cases, the Commissioning Agent may offer rental of their company-owned PQMs – provided their calibration certificate is current. For PQMs, supplier rental rates might be $300/day, $600/week, or $1.5k/month. For load banks from 5kW to 2500kW, rates might range from $350/day to over $15k/month. Costs for shipping, temporary cable, and accessories also need to be reviewed. When testing medium-voltage equipment, a step-down transformer might be needed to power a low-voltage load bank, which could be an additional cost possibly in the range of $1k to $10k/month. Roll-up standby generators or chillers might range from $5k to $15k/month. Do not forget to consider the additional cost for fuel when generator testing. The project site may also need to account for infrastructure costs to accommodate the temporary setup, and Installing Contractor labor hours must be considered as well.

Summary: renting test equipment is expensive. Make sure the project budget is clear on this scope. It is also important to balance performance expectations with a realistic budget. In particular, make sure the design specifications align with the anticipated testing, or clearly identify testing as a deferred cost until the Commissioning Agent defines it.

The best way to establish a test equipment budget is to request an official quote from a Test Equipment Supplier.

<u>Rent Versus Buy: Load Banks</u>

The Project Team should discuss standard maintenance cycles for equipment to determine if the Owner wants to rent or own a site load bank. If load banking is a common activity due to light overall building electrical load, purchasing a site load bank could be an overall lower cost. If load banking will be frequent or for multiple weeks of testing, Manufacturer case study summaries have shown less than a five-year payback for purchasing compared to renting. The payback could be even faster if the Design Engineer considers the building occupant or process load and only sizes the load bank to meet the minimum 30% diesel generator loading (to protect from wet-stacking). Natural gas generator sets are not subject to wet-stacking conditions.

Having a smaller permanent load bank for functional testing may be okay if building load is also available to equal the system full-load. However, the building load makes it difficult to provide true block-loads for transient response confirmations. It is also unlikely to maintain building load consistently for extended burn-in testing (over multiple hours). Following NFPA 110, the Day 1 testing may be difficult without having full load bank capacity for the commissioning process, either as a permanent or rental load bank. Purchasing a site load bank comes with additional costs, such as an equipment pad, external disconnect, and regular maintenance.

Rent Versus Buy: Power Quality Meters

It may make sense for the project to purchase portable PQMs instead of renting them. These would then become the property of the Owner and available for the Electrical Maintenance Team members. For PQMs and the necessary accessories, depending on the specific meter and features selected, the purchase cost may be between $5k to $15k each. The Owner then takes on the cost to have the PQM calibrated every twelve-months, replaces the internal batteries as needed, and provides firmware updates. PQM Manufacturers commonly provide a National Institute of Standards and Technology (NIST) traceable calibration certificate for $100 to $500 per meter.

Software Tools

Data Management

A significant portion of the commissioning scope involves managing data. The volume of data can be considerable. Examples of the types of data that needs to be managed:

- Checklists
- Test scripts
- Asset IDs/tags
- Discrepancy/Issues Logs
- Issue tracking (status and comments)
- Progress tracking of completed documents

Before database software tools, the industry utilized spreadsheets and plain text documents to manage the commissioning process. Some of the industry still relies on this approach, and it may be appropriate for smaller projects. With the manual method, test scripts are printed out and completed by pencil and paper. The Commissioning Agent then scans those handwritten copies for the final commissioning report, or sometimes keyboard inputs the data into the computer for a cleaner print. Optionally, a laptop may be brought to the site for keyboard entry during the tests.

Much of the commissioning documentation will have multiple revisions before being finalized. It is also very common to duplicate scripts in the event of re-tests. If the process is by manual entry, it is prudent to determine a document revision tracking method.

Mike Starr, PE

Cloud-Based Advantage

Fortunately, much of the industry is leaning away from manual processes and turning to software database platforms for managing commissioning documentation. There has been a shift away from the simple word processor and spreadsheet programs; although, there are some successful Commissioning Providers still using the manual approach but via cloud filesharing methods. By using cloud-hosted documentation, all Project Team members have visibility to view and edit the documents on-demand. For greater efficiency, and especially for projects of scale, cloud-hosted commissioning software packages are preferred. In addition to data management needs, this approach has several advantages, such as:

- Access to uploaded documents
- Picture taking throughout the process
- Commissioning document review type functions
- Management of Commissioning Team deliverables

Having the documents available via a cloud-based approach is becoming the standard practice because it allows on-site Contractors to seamlessly view and edit documents via portable electronic-tablets. Once the user inputs information into their local device, they synchronize to upload the data to the subscription-based online software. The Project Team sees live progress by logging into the software online, and the Commissioning Agent also benefits from efficient and quick report creation. For example, the Electrical Commissioning Agent may filter all MEP equipment results down to only the electrical equipment to see how many of the step-down transformers have completed Level 3 Installation Verification Checklists (IVCs).

Another advantage of using a software management approach is keeping commissioning conversations in a central place. Instead of managing a list of people that should be email cc-ed (carbon-copied) on commissioning emails, the software includes the ability to make comments and receive email alerts. Overall, software management solutions help the commissioning process be more transparent. For example, if the Commissioning Agent opens an issue, they can assign it to the Contractor responsible for the installation. The Contractor may then add comments to the issue as needed. Without commissioning software, these conversations normally occur through a series of email chains that are only viewable to the members copied on the emails.

Common Platforms

In deciding which commissioning software is best for your project, be sure to talk with your industry contacts to understand what has worked well for other people. Ask about the projects' scale and be sure to account for staff hours needed to develop expertise in the software platform, as the Commissioning Team usually takes on the responsibility of educating and troubleshooting software issues for the Owner, Design, and Construction Teams. In making your software choice, be mindful of any client concerns regarding the cloud-based nature of the data. Clients may have security precautions for which cloud service provider's servers their propriety information gets backed-up on.

Examples of cloud-based software options: CxAlloy, BIM360, Facility Grid, and Building Start. This list is not all-inclusive, as there are many other platform options on the market. As they are database-driven, most software includes functions to import spreadsheet data for quick test and checklist template creation.

Chapter 3: Commissioning Process

Project Phases

For new construction, in general, projects go through the following phases:

1. Pre-Design
2. Design
3. Construction: Levels 1-5
4. Maintenance/Operations

Existing project phases are only slightly different compared to new construction:

1. Planning: Update the Owner Project Requirement (OPR) to match the Current Facilities Requirement (CFR) documentation
2. Design
3. Install
4. Acceptance
5. Post-Acceptance

For more information, reference ASHRAE 0, Annex B: Commissioning Flow Chart.

The design and construction phases will depend on the construction delivery method. In a traditional design-bid-build approach, the project phases follow:

1. Concepts
2. Schematic Design (SD)
3. Design Development (DD)
   Value Engineering (VE)
4. Construction Documents (CD)
   Construction Administration (CA)
5. Post Construction

The American Institute of Architects (AIA) describes an alternate delivery method called the Integrated Project Delivery (IPD), which has been gaining popularity. Although they are not a good fit for every project, when the Owner has staff with strong technical knowledge about construction processes, IPDs may be very beneficial resulting in lower cost and shorter schedules. Using an IPD method involves much earlier involvement from the entire team rather than bringing on members only when it is time to complete their scope of work. The following are a few examples of IPD working relationships. Due to many various industry descriptions, these are summaries rather than official definitions.

- **Pre-Construction (Pre-Con) Services:** Before producing construction documents, Contractors are added to the Project Team for pricing and constructability assessments. These Contractors are typically required to bid competitively on the project after the Design Team establishes the construction plans and specifications. In some cases, they are precluded from bidding on the project.
- **Turn-Key:** The Installing Contractor employs in-house Engineers or engineering partnerships to design, construct, and quality check the installation from a single source contract; this may also be referred to as Engineer Procure Construct (EPC). If desired by the Owner for cost reduction, in some cases, turn-key Contractors can source used equipment. These Contractors may also sell post-construction Preventative Maintenance (PM) programs where they make visits to the building periodically to assess and up-keep systems.
- **Design-Build:** A consultant or Contractor leads a single contract team, similar to turn-key, but the Project Team seeks to maintain traditional project phasing. The Installing Contractor might also be the Engineer of Record (EOR), so the design documents may not need to be as detailed as a design-bid-build project. For example, to be more efficient with resources, an IPD may have a Design-Build Contractor complete the branch circuiting and panel schedules rather than the Design Team performing the work

The concept phase documentation could be useful to find an IPD Contractor partner. When possible, it is helpful for the Contractor Teams to be included at the design phase and awarded the construction scope up-front rather than having to competitively bid against other Contractors after the construction documents are sent out. Being an IPD Contractor partner is a time commitment. IPD design efforts include multiple pricing alternatives. Detailed BIM modeling of equipment/pathways is also common to confirm constructability. An IPD Contractor spends time asking their Vendors to provide sharp-pencil quotes and they are more committed to the design effort if they know their company will also provide the installation. Without the guarantee of the install scope of work, Contractors may be hesitant to participate in the IPD process fully. The Owner should work closely with the Design Team to determine if there is a suitable IPD arrangement for the project delivery.

Since there are more up-front team members in an IPD project approach, scope gaps are common; this is when assumptions are mistakenly made thinking a different project partner is responsible for a task or part of the installation. Of greater concern, this mis-coordination has impacts on schedule and cost. Scope gaps must be well managed to avoid significant pitfalls. The industry continues to find improvements with IPD processes by making team responsibility matrices.

Mike Starr, PE

Now that we have overviewed the traditional and modified project delivery methods, the following sections overview the traditional project phases for a new construction project. The discussion uses a commissioning emphasis and highlights electrical specific examples.

Pre-Design Phase

Pre-design aligns with the concept design phase. In this phase, the building is just starting to take shape. The Owner and Design Team are establishing building requirements and rough plan layouts. The industry encourages a knowledgeable Commissioning Team to be included at this stage. Being involved early in the design phase helps the Commissioning Team understand the Owner's expectations and overall design intent. Pre-design is an opportunity to baseline design intentions. Having an experienced Commissioning Team will bring forward their testing experiences and areas of concern for the Design and Owner Team to consider. Especially for complex applications, the project benefits from having the Commissioning Provider join the team early in the pre-design phase.

In practice, Project Teams often wait until the "For-Construction" documentation is available before hiring the Commissioning Provider, assuming that less involvement equates to lower costs for services. Caution: some certifications or standards, such as LEED and government buildings, require the Commissioning Team to be included at the beginning of the design. For less technical building applications, it may be acceptable to wait until after the construction documents portion of the project to hire the Commissioning Team. For instance, a low-rise (highest occupied floor less than 75-feet above grade) office building without specialty certifications may decide to wait until the construction phase if they plan to hire a Commissioning Team.

A front-end, lessons-learned meeting is highly recommended, especially when the Commissioning Team is included early in the design process. If the project has not yet started, what is there to learn? This meeting is an opportunity for all team members to offer suggestions based on their past project experiences. The lessons-learned may be regarding the building design at-hand or related to the administrative processes the team is about to partake in during the design and construction phases.

Design Phase

Commissioning Review

The Commissioning Agent's role in the design phase is to perform document reviews at key project landmarks. For example, 100% SD, 50% CD, 95% CD. The Commissioning Agent reviews documents for coordination and compares them to the Owner Project Requirements (OPR), as well as the Industry Standard of Care (codes and standards applied to modern equivalent building types). Reference NFPA 70B, Annex G for diagrams, and Annex F for symbols. NECA 100 is also a helpful reference for plan symbols. Consider there are distinct differences between ANSI and IEC electrical plan symbols.

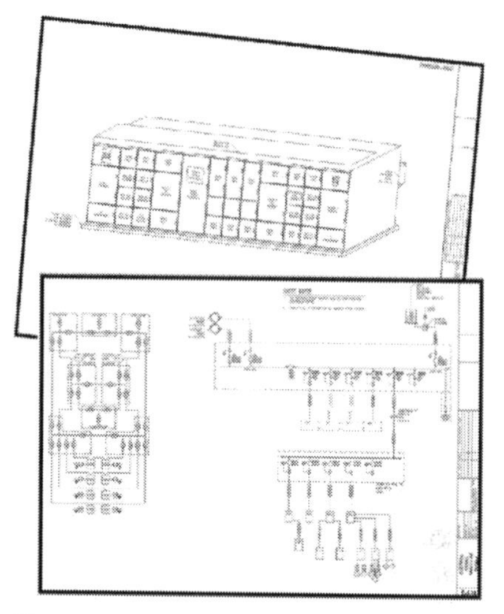

Figure 4: Electrical Plans

Once the equipment names are determined, establish site-specific asset IDs/tags for all equipment and systems to be commissioned. For example, panelboards might use building level and column numbers for names – this is determined by the Design Team. By setting up the equipment IDs for assets early in the design phase, document reviewers may associate their comments directly with the equipment IDs, ensuring everyone is referencing the same equipment. Design comments are then available as a record from early design through building turnover. In practice, the system designs are not matured enough for the Commissioning Agent to establish IDs until nearly the end of CDs.

A commissioning review is not the same as a peer, quality control, or code review. Those reviews are still the responsibility of the Engineer of Record (EOR). The limitation of an acceptable commissioning review comment is blurry – ask any group of professionals, and they will all have unique opinions about this. As the Commissioning Agent is not the Designer of the system, it is best to avoid comments that suggest a different approach to meet the same design intent.

Mike Starr, PE

The following sections suggest three types of comments that may be considered appropriate for a commissioning reviewer to make:

1. Verify the design meets the contract requirements
2. Confirm equipment and systems may be installed, tested-operated, maintained, and eventually replaced
3. Review the design for code and safety

Cx Review Consideration #1 of 3

Confirm the design shown on paper meets the contract requirements:

- **Owner Project Requirements (OPR)** or Request for Proposal (RFP). See ASHRAE 202, Annex D, for a suggested OPR format. Examples:
  - Exterior lighting may need shielding if the project is pursuing LEED credits for light pollution.
  - The Owner's insurance carrier may have indirect requirements. For example, Factory Mutual (FM) Global is a common building insurance carrier that allows free access to their datasheets via their website. Does the design include the FM Global acceptance testing requirements? https://www.fmglobal.com/research-and-resources/fm-global-data-sheets
  - The Owner may standardize on closed transition transfer equipment to better facilitate their maintenance programs.
  - Some facilities implement Power Utilization Effectiveness (PUE) dashboards and require target metrics at various load conditions.
- **Basis of Design (BOD).** This document is written by the Design Team to outline the criteria for their design. Depending on the review stage, the Commissioning Agent could be critiquing the BOD itself or comparing it against the produced plans. Examples:
  - Does the BOD consider the Owner's campus master plan and implement the client's standards?
  - Does the planned service approach consider the local utility requirements?
  - Has the utility fault contributions and X/R ratio been considered in the design?
  - Do the equipment types align with the Owner's current maintenance program? If not, are they willing to accept a new equipment type?
  - Does the project align with industry-specific codes and standards, such as NFPA 99 and Facility Guidelines Institute (FGI) for healthcare applications?
  - Does the design consider the type of construction occupancy and impacts related to the planned equipment? Reference NEC, Annex A (Product Standards), and Annex E (Construction Types).
  - If the BOD notes there will be remote annunciators for the UPSs, generators, or transfer switches, are they located on the floor plans?

Cx Review Consideration #2 of 3

When reviewing the project plans and specifications, can the equipment/systems be…

- **…installed?** These may be physical constraints or coordination items with the related building systems. Examples:
    - Is the mechanical system coordinated with the liquid-cooled rotary UPS?
    - Do the documents identify who is responsible for programming the relay configuration files? Is it the power system study author, or do the design documents clearly explain the desired functions for the Contractor to build the relay database files?
    - After the initial installation is complete, is there a workable option to install future equipment?
    - In an existing space, does the equipment fit, including required clearances? Will the equipment need to become tilted to an extreme angle to fit through existing openings? If yes, is the Manufacturer aware so they may design additional internal component bracing?

- **…tested-operated?** Examples:
    - Is this a modular design that benefits from built-in test features?
    - Are there any concerns for transformer Transient Recovery Voltage (TRV) or ferroresonance?
    - Is the design compatible with the requirements listed in the Commissioning Plan and specifications?
    - Will the project be renting rack mount load banks, or does the plug-in busway need to be a distance off the wall for a temporary end-tap box?
    - Do the cooling system controls need to be on UPS power to ride-through generator starting, or does the sequence allow those controls to lose power and restart?
    - Even though it is not a code requirement, especially for mission-critical high-density electrical loading, have duct bank heating calculations been performed to confirm duct bank ampacity?

- **...maintained?** Examples:
    - The code clearances for the pad mount equipment are met, but will the operator have room to use a hot stick with the available clear space?
    - Does the site electrical equipment located near driveways have traffic barrier protections?
    - Is exterior electrical equipment with only heater-thermostat combinations sufficient, or given the project location, is a humidistat more appropriate?
    - As the project can afford, does the design include Preventative Maintenance (PM) enabling features? For example: Infrared (IR) viewing windows, remote annunciators, mimic bus on the face of sophisticated equipment, partial discharge sensors, arc flash reduction means, arc-resistant gear, capacitively coupled live voltage indication, conduits color-coded based on system, integral control or energy management/reporting type circuit breakers, integral thermal monitoring for trending termination temperatures, external transformer sight gauges, etc. Implementing safety features may reduce the likelihood of an electrical hazard occurring (lower risk).
    - Does the design consider safety-related design criteria identified in NFPA 70E, Annex O?

- **...eventually be replaced?** Examples:
    - Will the large equipment be able to be replaced through the room's doors? Or should the design incorporate other options, such as a structural knock-out wall? Is the building path rated for the equipment weight, such as the raised floor, elevator, etc.?
    - Will the building be allowed to experience an extended power outage, or is the Owner better served to add redundancy in the current design to avoid a future outage during replacements?
    - Is there ample space to revise equipment layouts or system configuration? For example, upon end-of-life, if a parallel UPS system changes from a centralized static switch to a distributed design, the feeder lengths of all parallel UPSs must be identical for equal impedance during static bypass mode. If there is not enough floor space in the building, this may create an installation challenge and/or required outages in the future.

Cx Review Consideration #3 of 3

Does the design have a code or life safety concern?

To clarify, a code/life-safety review is not explicitly in the Commissioning Team's scope of work. Formal commissioning programs note that code or regulatory type reviews are outside the base scope. However, as licensed professionals, code/life-safety type comments should be applicable for a commissioning review. These comments are not a substitute for the EOR's code review.

Examples:

- Do emergency feeders require two-hour rated systems?
- Are the electrical room walls rated, and do they need to be?
- Is the modular structure subject to NEC 646 or UL 60950?
- Are OSHA mandates, such as eyewash stations, applicable?
- Based on the voltage and type, is the indoor transformer required to be in a vault construction?
- Is the exterior generator far enough away from the building, and does the exhaust stack height comply with local requirements? Is the generator exhaust the code distance away from ventilation air intakes?
- Is there a building two-way communication system, and does it require emergency back-up power? If yes, for how long?
- Upon loss of normal power, will the project meet the building code-required emergency lighting levels?
- Does the local jurisdiction approve of natural gas for emergency generators?
- Does the service arrangement, per NEC, require a campus switching procedure?
- Is the maximum allowance of fuel storage for the occupancy type considered?
- Do all electrical system components meet the system Basic Insulation Level (BIL) requirements?
- Will the emergency generator scheme power the fire pump for the NFPA 20 required eight-hours?
- Do the design documents describe the cable shield bonding requirements (single-point, multi-point, cross-bonded)?
- Are exits signs visible within one hundred feet, per NFPA 101 (if applicable for the jurisdiction)? Has the code adopted the approach of dynamic signs?
- Are there energy storage systems, and do they meet the current codes based on their storage type, capacities, and product listings? Is a separate battery room required? Reference: NFPA 855, NFPA 1, NFPA 70E, and/or IFC.

Mike Starr, PE

- Is the data center utilizing NEC, Article 645? If yes, does it meet the stringent requirements? For example, a design may use this article to allow a rack power strip to plug into a receptacle beneath a raised floor.

Cx Review Versus Peer Review

Focusing the commissioning review on these three areas (noted in sections above) is a broad review scope. Unless the review is a peer review, the reviewer should accept the design put on paper by others. The goal is to determine if the system, as designed, may be commissioned to meet the design intent.

If the Commissioning Agent's experience also includes design of electrical systems, the Owner may consider requesting a more detailed peer review of the design documents. It is more appropriate to recommend alternate designs in key areas during peer reviews, provided with explanations, for the EOR's consideration. Within reason, the peer reviewer may also suggest alternative approaches to documenting the electrical scope; for instance, they may have seen past project coordination success if the Design Team provides an integration points list on plans or within specifications.

Many Project Teams choose to conduct document review sessions via a software package. Such applications may include file-sharing features for multiple users to contribute mark-ups to central documents directly through a desktop application. If you are willing to take on a small amount of project coordination to set it up correctly, Bluebeam Studio or similar software may prove to be a useful PDF tool for virtual group mark-up sessions.

Upon receiving either a commissioning or peer review, the EOR usually reviews and responds to the commissioning comments. If contracted, the Commissioning Team may do a follow-up review of the Design Engineer's updated project documentation once available.

<u>Value Engineering / Value Enhancement (VE)</u>

If invited, the Commissioning Team should be an active contributor for VE sessions. Simply put, VE is a process to reduce the projected cost for construction. Projects may not decide to conduct a VE process unless the project budget is at risk of being exceeded. Even if the project does not have a budget challenge, a short-version VE process is recommended. If the estimated construction cost is excessively over budget, the VE phase may also identify significant scope reductions. For example, entire sections of electrical distribution may change to "future," with only conduits and pull strings installed on Day 1. A Design Team generated cost estimate typically yields conservative savings. In some cases, the

Owner and Design Team may decide to request a cost estimate from a local Contractor or Construction Manager (CM).

Like commissioning, VE is a process rather than a moment in the design phase. Ideally, for the sake of minimal Design Team re-work, VE happens in the DD phase rather than the CD phase. The VE process may take one or more weeks to complete, as team members usually reach out to Vendors for pricing accuracy. During VE, consultants and possibly Contractors brainstorm and reconsider the system design approach. Contributors develop lists and assign a cost savings amount for each consideration. Then, with a knowledgeable Project Team to discuss operations and maintenance, the stakeholders weigh the proposed VE items needed to stay within budget versus the resultant long-term impacts of those decisions. After the VE process, the Design Team spends time updating documents to match the accepted VE items.

VE can be a valuable process, not just a cost reduction strategy. For example, the VE phase may create a conversation with an Owner who did not have the background to understand resiliency and maintenance differences between switchgear (UL 1558) and switchboard (UL 891) construction. By establishing a conversation during VE, the Owner discusses options with their staff. They may determine the less expensive switchboard construction approach is most appropriate for their project; equally, they may determine switchgear construction is required. Further, the Owner Team is then an informed contributor to the design decisions.

Commissioning Documentation

Once a design has been established, and the commissioning review has occurred, the Commissioning Team forms installation checklists, coordination plans, and test scripts (further details on these in upcoming chapters). Every team member sees the installation from a different perspective and has unique contributions: Electrical Contractor, Controls Contractor, General Contractor, Construction Managers, EOR, the Manufacturer Service Representative who will be doing the start-up, Owner Maintenance Staff, etc. As such, a team review of the commissioning documents is ideal. By collaborating on the approach, the test process is massaged on paper rather than on the day of testing.

The Commissioning Team usually issues commissioning specifications. The Design Team may include commissioning aspects in their design specifications, but it may be easier to coordinate if the Commissioning Agent provides this definition instead. The Construction Specification Institute (CSI) has assigned commissioning to Master Format Section 01 91 00 (Division 1).

Mike Starr, PE

For an example Division 1 commissioning specification, reference:

https://www.wbdg.org/ffc/epa/federal-green-construction-guide-specifiers/01-91-00

Many Commissioning Providers also include division specifications for each building system. For example, electrical commissioning may issue a Division 26 commissioning specification.

The CSI specification divisions are generally as follows:

- Division 00: Procurement and Contracting Requirements
- Division 01: General Requirements
- Division 02 through 14: Facility Construction
  - Architectural, Interiors, Structural
- Division 20 through 29: Facility Services
  - Mechanical, Electrical, Piping, Technology, Fire Protection, Instrumentation and Controls
- Division 30 through 39: Site and Infrastructure
  - Civil
- Division 40 through 49: Process

The Commissioning Plan (discussed in the Project Management chapter) focuses on roles, process, scope, and schedule; whereas, the commissioning specifications provide more detailed requirements, such as:

- Defining non-conformance metrics and how testing will address deficiencies
- Identifying intermediate sign-off stages and which project role is responsible for said sign-offs as the project progresses
- Requiring torque logs to be submitted for the project record
  - Manufacture torque recommendations should be used. In absence of manufacturer direction, NEC Annex I (UL 486A-B) or NETA.
  - Reference bolt torque calculator based on bolt type: https://www.portlandbolt.com/technical/bolt-torque-chart/
- Adding acceptance testing to the commissioning specifications – if requested by the Owner and the Design Engineer did not already specify

Construction Phase

As mentioned in the pre-design section, the construction phase is often when the Project Team decides to contract a Commissioning Provider. Ideally, the Commissioning Team joins at the start of the project, but this does not always happen for various reasons. If this is their introduction to the project, the Commissioning Team will need to do a modified review of the contract documents to come up-to-speed with the project scope. In this review, the Commissioning Agent makes pre-construction comments for the Project Team to consider.

Construction Administration (CA)

If contracted, a common arrangement for this stage of the project is having the Commissioning Provider engaged in the CA process. The Commissioning Agent reviews Contractor submittals/shop drawings and Requests for Information (RFIs). The Commissioning Team may also issue RFIs to the EOR if they have formal questions. Still – the EOR remains the primary reviewer for these documents. Ultimately, the EOR and Contractor are responsible for utilizing the CA process to agree that the planned installation meets the design intent. Note that the EOR's review does not relieve the Contractor from adhering to the requirements on the contract documents (plans/specifications).

The same rules used for the document reviews in the design phase apply to the commissioning review of the Contractor submittals and RFIs. In particular, the Commissioning Agent is looking for documentation that may be required beyond what has been called for by the EOR. For example, the EOR may have experience with the Contractor's selected equipment and not be concerned about seeing interconnection wiring diagrams to approve a submittal. The Commissioning Agent should consider the installation steps and comment on the need for the Contractor to submit control wiring diagrams. Another example: the Commissioning Agent may comment on a switchboard submittal to include Time-Current-Curves (TCCs) for all circuit breakers, as the Contractor Team will need the TCCs for the power system studies.

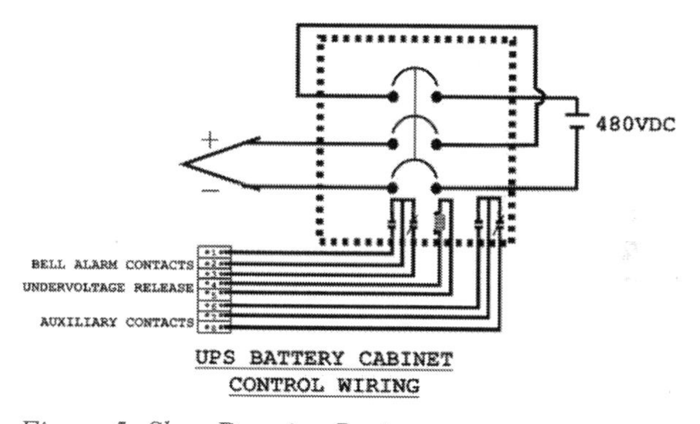

Figure 5: Shop Drawing Review

"Submittal parties" are sometimes scheduled to help provide a group review of the submittal information, wherein all project stakeholders and the product Manufacturer meet to review major equipment submittals together. Given the opportunity, the Commissioning Agent can be a useful team member to join the review party. Some might see submittal parties as a way to avoid or rush individual reviews, but this mindset is discouraged. The process works best if all team members are given advance time before the meeting to perform their detailed review. The advantage of submittal parties, which does lead to time savings, is on the back of the process, as the group review and spoken conversations mitigate multiple Contractor re-submittals. The team has discussions and arrive at resolutions faster than they would by exchanging comments through the normal written submittal process.

Commissioning Kick-Off

The construction phase is when the Commissioning Team Lead initiates the project commissioning kick-off. Commissioning meetings are then re-occurring. The frequency of the meetings increases as the testing scope begins. In these meetings, the Commissioning Plan, Test Equipment Plan, management software, and commissioning specifications should be reviewed as a group to see if there are questions or areas for further collaboration. The commissioning meetings should also focus on the anticipated Owner training scope. Having training conversations helps the Owner think ahead about which staff needs to participate. At this point, if required for larger projects, the Contractors leading the commissioning scope for their trade submit Quality Plans for the Commissioning Agent to review and comment.

Equipment Inspections

Before equipment inspections, perform live-dead-live procedures and Lock-Out-Tag-Out (LOTO). Inspections may occur in factories or at intermediate stages during the construction process; thus, it is critical to confirm safety twice and always check for multiple power sources. For example, it is common for switchgear to include separate branch circuiting from station batteries used for control power. See the Safety chapter for further discussion.

During electrical equipment inspections, depending on where the equipment is (at the Manufacturer's facility versus final installation location on site), consider these types of activities and checks:

- Photograph all equipment nameplates
- Doors should operate freely without jamming
- Control and relay wire management is orderly
- Protection settings meet the project requirements

- Installation is in accordance with Part 3 of the design specifications
- Equipment code and ventilation clearances meet or exceed requirements
- Communication cards are provided where equipment specifications require them
- Acceptance testing decals have been placed on the individual components that required testing
- Before energization, the equipment has been cleaned – free of dust, packing materials, and other debris
- Feeders include cable lashing if required by the Engineer or Manufacturer (usually due to equipment short circuit ratings in excess of 65kAIC)
- Bus bars have labels on them, so the operator may identify the phases (A, B, C) when infrared scanning occurs
- Physical layout and ratings are compatible with the engineered design and match the product shop drawings
- Wires are secure and seated properly. Wire pulling (spot checking control wire connections) is a debatably beneficial measure.
    - If the team finds wire pulls to be appropriate, gently push control/signal wires clockwise to avoid loosening screws on already tight connections.
    - The Manufacturer should be transparent with torque values used for the various wire terminations. Some third-party validation teams will inspect control connections one-by-one, using small calibrated torque screwdrivers: https://www.checkline.com/product/DID
- All control compartments are serviceable. For example, rack-able breaker rails may overlap control terminal blocks, making the control wire connections unserviceable while the breaker is racked-in. Or maybe the control compartments are full of wiring and would benefit from built-in lighting.
- Torque marks are un-broken on all factory connections; likewise, at the final installation location, Contractor terminations have un-broken torque marks.
    - Consider specifying a requirement for Contractors to utilize torque seal, not just a black marker. Many other industries use torque seal, and the electrical industry is starting to see the benefits also.

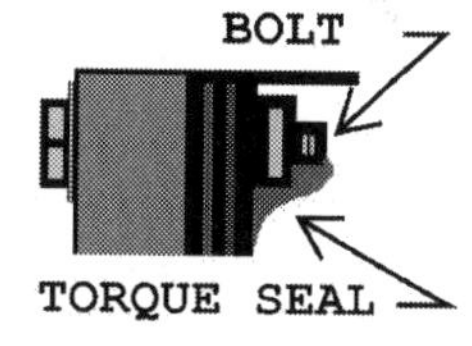

Figure 6: Torque Seal Application

- Fuse legends and Sequence of Operations (SOOs) are posted on or near the equipment
- Wiring crossing sharp equipment edges have a rubber guard or other protection applied to shield the conductor insulation
- Interior control components have labels at terminal blocks and they match the control shop drawings to promote serviceability
- Equipment bonding is provided from section-to-section and at all equipment doors that have mounted electrical components and wiring

Mike Starr, PE

Reference testguy.net for a wealth of information on equipment inspections. The example below is for switchgear and switchboards. In navigating the website, many more insights are available, including acceptance testing references.

https://testguy.net/content/258-Switchgear-and-Switchboard-Inspection-and-Testing-Guide

Field Observations

The primary use of the Commission Agent's time during site walks focuses on the systems they have been contracted to commission. Still, the Commissioning Agent is a walking-talking quality assurance measure there on behalf of the Owner Team. By performing field observations and documenting discrepancies throughout construction, the Commissioning Agent affords the Contractor Team more time to address installation issues. Site policies should be reviewed before taking photos in case photography is not allowed or the subject material is limited. The following are examples of recommendations for field observations of electrical systems (subject to code and design specifications):

- The installation provides identification for electrical systems
    - Tape colors match site premises wiring colors
    - Data center whitespace includes wall lettering if the ANSI/TIA 942 criteria or similar is required
    - Service entrance equipment includes labeling for the rotation: CCW or CW (Counter-Clock-Wise, or Clock-Wise). See the Phase Sequence, Rotation, and Arrangement section for further insights.
    - Equipment nameplates from the Manufacturer are detailed, and any site-specific Contractor labels conform to the project requirements
        - Arc flash labels are located at equipment main disconnects only, all gear sections, and/or front back of the equipment, as required.
        - The Owner may consider an add-service for the Commissioning Team to provide a customer site Labeling Guide, which involves reviewing the design specifications and Contractor submittal data. A comprehensive review of this Labeling Guide by the EOR is likely needed, as typically design documentation does not address every labeling instance. Having the Commissioning Team create a Labeling Guide is an opportunity for clarity.
- Poured concrete equipment pads are level. Channel leveling and/or structural details are followed if required by the project documents.
- Electrical distribution equipment includes as-built panel schedules
- A printed copy of the power system study is available for reference
- Design specification Part 1 sections, which describe job-site material handling procedures, should be reviewed for field conformance
    - For example, cable reels should not rest on their flat surface (side) or risk damage to cables due to other cables' weight.

- Grounding and bonding connections have been assessed before they are poured in concrete or covered by earth
- Fire-rated walls have fire stopping at penetrations and labeling. The UL system used is in agreement with what has been installed.
- Network connections at the equipment requiring data signal for point-to-point and graphics confirmations have been installed
    - For example, connectivity for a UPS network card is needed to confirm the software points in the BMS.
- Uni-strut trapeze supports hang from building structure rather than other uni-strut hangers unless calculations have been performed and this is permitted by the design specifications

The Commissioning Agent should follow-up every site visit that has site observations with a detailed report and updated Issues Log. Ideally, the report also includes general comments about the state of construction, such as how far along the installation is, areas that are complete or have not started, and systems that are in the process of being installed.

Consider reviewing these helpful references for further information on field observations:

- NECA 1: Good Workmanship in Electrical Construction
- NECA: Equipment and material standards 100 through 700
- ASHRAE 202, Annex J: Construction Observations and Test Checklists and Reports
- NFPA 70B
    - Annex E: Suggestions for Inclusion in a Walk-Through Inspection Checklist
    - Annex M: Equipment Storage and Maintenance During Construction

Issues Log

Also referred to as a discrepancy log or Corrective Issues Report (CIR), the Issues Log is an in-progress, multi-discipline summary of all commissioning related deficiencies found through the process. This log usually captures the specific resolutions for issues (Issues and Resolutions Log). For an easy understanding of concerns, each issue in the log should include a reference to the contract documents or presiding standard/code.

See ASHRAE 202, Annex K, for an example of an issues and resolution log. The log includes options to track the status of issues. At any given point in the process, issues may change states (open, in-progress, closed). The goal is for all issues to be resolved and closed promptly. Critical issues impacting cost or schedule may be marked and reviewed more frequently. When tracked via manual spreadsheet, the Commissioning Agent typically makes regular updates to this document and sends it electronically to the entire Owner and Project Team. The Contractor Team then works to address each item in the Issues Log – this is accomplished by updating the installation to meet the requirements and/or reaching out to their suppliers to address deficiencies.

Cloud-platform commissioning programs include an Issues Log as a standard function. They may also include the ability to assign issues by the equipment asset IDs and automatically notify the responsible Contractor by email. Logs that include photos of the discrepancy help avoid the additional time walking the job site to see the concern in-person. Commonly, the individual addressing the install issue takes a photo to upload as proof before they mark the issue as resolved. The Commissioning Agent monitors resolved issues and closes them once verified. Issues may also be assigned to the Design Team for a response if necessary.

As-Builts and Record Drawings

Upon substantial completion, the Contractor Team will typically submit as-built drawing mark-ups to the Design Engineer. If contracted, the Design Engineer then forms final record documents. Owners might consider an add-service to have the Commissioning Team review the record drawings to confirm the building has accurate final documentation.

Commissioning Levels 1-5

Most prevalent in the mission-critical market sector, but certainly applicable to other markets, Owners determining desired scope and providers writing proposals use Commissioning Levels 1-5 to describe commissioning. The levels build off one another, with Level 1 being the entry point into the process and Level 5 covering all commissioning aspects, including Levels 1 through 4. An Owner may request Level 5 commissioning and exclude certain activities as desired. For example, Level 1 factor testing could be removed from the scope, but Level 5 Integrated System Testing may still be performed. These levels segment the project's construction phase:

Level 1: Factory Acceptance Testing (FAT) / Integrator Acceptance Testing (IAT)
- Factory Witness Testing (FWT) / Integrator Witness Testing (IWT)

Level 2: Site Acceptance Testing (SAT)

Level 3: Installation Verification Checklists (IVCs)
- Equipment Start-Up

Level 4: Functional Performance Testing (FPT)

Level 5: Integrated System Test (IST)

Level 1 – Factory or Integrator Testing

The commissioning level system starts at the equipment Manufacturer's facility. If required by the Design Engineer's specifications, the equipment may have a factory testing scope before shipping to the job site. This testing is Factory Acceptance Testing (FAT) or if a Project Integrator is on the team combining multiple pieces of equipment for grouped systems, Integrator Acceptance Testing (IAT). The rigor of factory testing may be visual/mechanical/electrical or even functional. If the Owner and Project Team want to witness the functional testing, the Engineer may specify this witness requirement in the design documents. The Contractor will have included the cost for a given number of people to participate in Factory Witness Testing (FWT) or Integrator Witness Test (IWT). Manufacturers are also starting to offer a Remote Factor Acceptance Testing (RFAT) option.

In a VE session, when the Owner is trying to reduce the projected construction cost, Level 1 testing is a prime candidate to remove from the project scope. However, in some cases, it is recommended to keep the FWT scope, as the best place for problems with the equipment or programmed sequence to surface is at the factory before shipping. This provides the least cost/schedule impact opportunity for resolutions.

Witness testing provides an opportunity for the Installing Contractor to see the equipment in-person. That Contractor is then able to inspect the Manufacturer's temporary installation to verify the anticipated connections and consider if the project site is ready to receive the equipment. Factory testing presents an opportunity for the Owner's Representative(s) to suggest functional changes to the equipment's operation to better meet the project needs. The factory testing also establishes baseline test data in a quality-controlled setting. If the equipment is not performing the same once installed on-site, the FWT documentation is a helpful troubleshooting reference. Manufacturers also offer environments to attempt tests that may not be repeatable on-site. For example, to confirm a UPS inverter protects itself from high temperatures, the factory may simulate the elevated temperature by changing a value in software – that may not be easily accomplished by the field technician. Bonus: the Manufacturer usually has accommodations that involve group dinner and/or other team-building activities. Allowing the Owner and entire Project Team to enhance their professional relationships will positively impact the project as-a-whole.

Which electrical equipment should have FWT? UPS, generator, automated switchgear assemblies, and prefabricated construction methods are all candidates to consider FWT. If the project is weighing the value of including FWT, think about whether the equipment is non-standard or if it is a packaged design readily available from the Manufacturer. Non-standard products benefit from more stringent testing. If the job includes multiple iterations of the same equipment or integrated system, performing FWT for at least the first unit establishes Owner and Engineer expectations to the Manufacturer for the remaining duplicate units.

Upon request, the Manufacturer might provide the planned factory test procedure. Given the opportunity, the Commissioning Agent should provide comments to see if the Manufacturer can accommodate any suggestions. In particular, comment if the Manufacturer test procedure does not test all operational scenarios. For example, what is the system response if the equipment loses control power or network connectivity? Another example: what will the system do if X, Y, or Z does not work in the middle of a sequence (failure modes), such as when a breaker fails to close when prompted?

Use caution while testing in the Manufacturer's facility. The entire installation is a temporary setup, so safety concerns are a priority. Life-dead-live and Lock-Out-Tag-Out (LOTO) procedures should be implemented. Even simple de-energized equipment inspections may benefit from a bump-cap style hard hat and eye protection. See the Safety chapter for additional considerations.

Upon arriving at the Manufacturer or Integrator factory, after the safety program requirements, the first thing everyone wants to do is turn the power on and see the functional aspects of the equipment in-action. However, the factory visit is the opportunity to comb through the equipment to confirm it is correct for the project. Factory equipment inspections are a quality step in the process; time and project funds have been allotted to ensure the product meets the project requirements. After lock-out and confirming the equipment is de-energized, before functional testing, the gear should be thoroughly inspected. It is not uncommon to spend one or more hours inspecting a single piece of equipment before energization. The Commissioning Agent should track issues found during this stage and includes them in the Issues Log. Ideally, the Manufacturer resolves the issues before beginning the functional testing part of the factory visit. Schedule permitting, the issues found during factory testing should be resolved prior to shipment.

Medium-voltage equipment usually proceeds with factory functional testing using low-voltage control power (125VDC or less). On the job site, after confirming functionality with control power, medium-voltage line voltage testing may begin. Systems operating below 1000V commonly use their true line-voltage for factory and site functional testing.

Level 2 – Site Acceptance Testing (SAT)

In Level 2, the equipment arrives at the project site to be installed by the Contractor. Once installed, depending on the project requirements, the Contractor or another entity may have Site Acceptance Testing (SAT) to perform (per specifications). Acceptance testing is completed while the equipment is de-energized and with safety grounds installed.

The acceptance testing for Level 2 commissioning activities may be witnessed, or the Commissioning Agent may just review the final test data, which comes from the tester with their interpretation of the results. At this stage, the Owner Maintenance Staff is encouraged to start maintenance logs and post them on the equipment. As the maintenance logs fill up over the equipment's life, the staff typically scans them into an electronic format for future reference.

Even if the building electrical system does not go through the rigors of a commissioning process, having electrical acceptance testing by a National Electrical Testing Association (NETA) or a National Institute for Certification in Engineering Technologies (NICET) entity is strongly encouraged for all electrical installations. The certifications held by these technicians may also have experience levels, such as Level 1, Level 2, and Level 3. The methods utilized align with testing recommendations based on IEEE, ICEA, NEMA, NFPA, OSHA, ASTM, ANSI, UL, and similar industry references. Equipment that has been acceptance tested is documented via a test report. The tester provides a decal sticker directly on the equipment that identifies their company information and the test date. See NETA standards documentation for test reports and testing decal labeling/color requirements. To locate a NETA certified company, reference their search tool (select the state from the drop-down and click query):

https://neta.netaworld.org/netassa/censsacustlkup.query_page

Do not maintenance test existing installations unless there is a replacement plan for failed test results (planned outage). For example, VLF withstand testing is a go or no-go test. Reference NETA and IEEE 400.2. If a cable is defective, it should be replaced or repaired before re-energizing.

Level 3 Through 5 Overview

These remaining commissioning levels involve scripts and checklists, which have an entire chapter dedicated to them titled: Test Scripts and Checklists. Here we will overview the levels and detail them further in the later chapter.

- **Level 3:** Installation Verification Checklists (IVCs) and Equipment Start-Up
  - Also known as pre-functional checks, System Readiness Checklists (SRCs), or Verification Test Procedures (VTPs)
  - Leading up to the equipment start-up period, the Installing Contractor completes IVC checklists developed by the Commissioning Agent. Every piece of equipment to be commissioned has an IVC. A single project may have hundreds of IVCs for the Installing Contractor to complete. The purpose is to verify the installation and create a more efficient handover of project data. The Owner is paying extra for the entire Project Team to be diligent in this scope of work.
- **Level 4:** Functional Performance Testing (FPT)
  - Also known as Functional Performance Procedures (FPP)
  - Individual equipment and systems are tested in all modes of operation. For example, the generator and transfer equipment will be functionally tested separately and then as a combined system.
  - There are fewer FPT scripts compared to IVCs on projects, as not all electrical equipment has functional characteristics.
- **Level 5:** Integrated System Test (IST)
  - Also known as pull-the-plug, lights-out, or noted by NFPA as black-start testing
  - Effectively, for the entire building commissioning approach, Level 5 is a multi-discipline test used to confirm the building operating conditions: all building systems working together (mechanical, electrical, controls, etc.)

Post Construction

<u>Owner Training</u>

As described in the Introduction chapter, one of the most significant benefits of providing commissioning on projects is enhanced Owner training. Ideally, the Contractors submit their training plans, and the Commissioning Team members review and comment. It is helpful if the Commissioning Plan or design specifications requires the Vendor or Contractor performing the training to have received instruction directly from the Manufacturer – this may require multiple training instructors and/or sessions. The training content should include inspection, testing, operation, and maintenance procedures for at least the primary equipment. Training will not necessarily provide in-depth instruction on maintenance testing methods, but should at least overview the frequency and types of tests appropriate for standard equipment life.

The Owner Team may need to receive approval from the Manufacturer ahead of time if they would like to video record the training, or the Design Team may write this requirement into the project specifications. All of the Owner's Maintenance Staff members likely benefit from training. Multiple training sessions may be necessary to accommodate all persons. See ASHRAE 0, Appendix P for an example training syllabus.

Although it is tempting to have everyone meet in the equipment rooms, it is recommended to have a classroom portion for the first half of the training. Equipment rooms are noisy. Providing a classroom setting provides the best opportunity for everyone to understand the system before they stand in-front of it. If the Maintenance Staff has not participated in the design or construction, consider sending out the training material ahead of the scheduled training.

During the Owner training, it is helpful to use a feedback mechanism to confirm everyone understands the installation. For example, a verbal quiz method: "How would we go about putting the system into bypass?"

Commonly, the Owner Maintenance Staff members are experienced in the system types but have technical knowledge gaps aided by this training. In many cases, the training helps Building Operators become familiar with a new Manufacturer for equipment that they already have a working knowledge of the primary functions. If possible, conduct the training before the building is serving critical load, so the operators may practice putting the equipment in the various operating modes without impacting regular business activity.

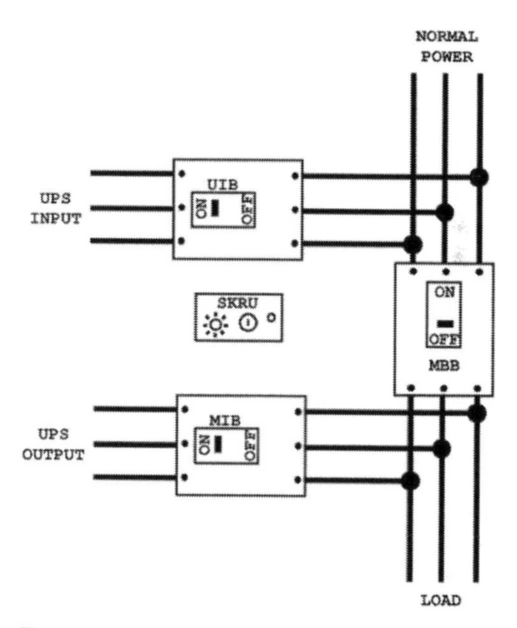

Figure 7: UPS 3-Breaker Bypass

Lessons-Learned

In the pre-design (concept) phase, the project hopefully started by reviewing past lessons-learned from other projects. A close-out lessons-learned meeting is also suggested. The Design and Construction Teams may even consider this to be a deliverable rather than an optional requirement. In some cases, Contractors also invite their Sub-Contractors and Vendors to the lessons-learned meeting. Everyone has lessons-learned that should be documented for all providers to benefit from. The Owner may even look back at these lessons if they have future buildings or expansions. Likewise, the Project Team may reflect on these lessons-learned for the design of future buildings. In working with the Owner, the lessons-learned process may be an opportunity to ask if the Owner is willing to share their energy metrics over time. Sharing metrics like this is incredibly useful to the industry for comparison to the design assumptions and may also highlight an opportunity for re-commissioning the building in the future.

System Manuals

System Manuals are a common add-service to a commissioning scope. See ASHRAE 0-Annex O and ASHRAE 202-Annex L for System Manual samples. Essentially, a System Manual is a focused Operation and Maintenance (O&M) manual. The Contractor delivers O&M manuals at the end of the project. An O&M manual is a substantial document (multiple three-ring binders) containing Manufacturer data for all system components. A System Manual should be condensed containing only the most commonly accessed data for day-to-day maintenance and operations. For example, a System Manual may isolate the Sequence of Operation (SOO) for all equipment. An operator would see how to put the system on utility Source-B without hunting through replacement parts product data. System Manuals also group critical maintenance information. For example, requiring fluid sampling for an existing pad mount transformer may help baseline its health over time. Commissioning Teams may even enhance these manuals by including custom system block diagrams that combine the engineered system diagrams with the specific equipment's Manufacturer data.

Final Commissioning Report

After testing, the Commissioning Team owes the project a final commissioning report. The report compiles all commissioning documentation for the project record. It also describes the test equipment, provides high-level takeaways from the testing, and summarizes issues that are not yet closed.

Admittedly, these are sizable reports that take time to assemble. For example, the Electrical Commissioning Team member has significant time investments to evaluate the PQM data; likewise, the Mechanical Commissioning Team has sensor data readings to extract, annotate, and write observations on. The final commissioning report may take between four to eight-weeks to complete after the project close-out. It would be prudent for the Owner Team to request scope for the Commissioning Team to have a review meeting - instead of blindly accepting the final commissioning report without being guided through the significant depth of its contents. Since this report is extensive, the Commissioning Team should add electronic bookmarks within the PDF file for easy navigation, or section markers, page numbers, and table of contents for printed materials.

Maintenance

The commissioning conversation would not be complete without highlighting maintenance. Through experience, Electrical Commissioning Agents become a great resource to discuss maintenance considerations. They bring value to the discussion because they are regularly commissioning new buildings with the latest technologies.

Still, there is an entire industry of operations, maintenance, and testing staff dedicated to maintaining existing systems. Given all building systems they have experience with (not just electrical), these Maintenance Staff members usually have a wide range of technical knowledge. Together, these industry experts see the electrical system through twenty to fifty-year life-cycles and have valuable experiences to share. With this in-mind, the following sections consider maintenance from a commissioning perspective.

Maintenance Programs

The commissioning process is an opportunity for Owners to review their maintenance programs. For an add-service, the Owner may request the Commissioning Team to participate in creating or updating an existing Preventative Maintenance (PM) program. Reference NFPA 70B-Chapter 4 and 5 for PM. Reference NFPA 70B-Chapter 30 and Annex N for reliability-centered maintenance considerations. For example, consider writing the calendar date installed on air/fuel/oil filters and batteries.

Documentation for comprehensive building facility management may include:

- **Standard Operating Procedure (SOP)**
  - Also known as Standard Work Practice (SWP)
  - This document describes the operation, maintenance, and other technical data for a specific piece of equipment. An SOP is somewhat of a cheat sheet, summarizing the most important maintenance and operating procedure steps. SOPs should incorporate Owner standards. For example, there may be specific safety procedures for cold-weather switching of less flammable type fluid transformers.
  - All equipment types have an individual SOP, which is referenced by Maintenance Staff when interacting with the equipment. These are commonly placed in protective plastic document sleeves and located with the equipment. The SOP usually includes photos and call-outs for quick user reference.

- **Method of Procedure (MOP)**
  - This document is a template for procedures. To make sure all required parties agree about the procedure, MOPs are very detailed and have multiple sign-off requirements. A MOP document is stand-alone and references the SOP documents to provide further context (procedural format). It is helpful to include a template of the MOP within the System Manual.
  - A Contractor or Owner Maintenance Staff member fills out a MOP before any significant maintenance or equipment state changes to the systems.
  - Reference MOP sample from The Up-Time Institute: https://journal.uptimeinstitute.com/the-making-of-a-good-method-of-procedure/
- **Current Facilities Requirement (CFR)**
  - Also known as Utility Management Plan (UMP)
  - This document caters to the specific building or campus and might be a combination of System Manual and system re-configurations that the Maintenance Team accumulates over time (historical logs).

Maintenance Frequency

The electrical industry is fortunate to have excellent operations and maintenance references. NECA, IEEE 902, NFPA 70B, NFPA 70E-Chapter 2, NETA, and Manufacturer product manuals are go-to references to develop facility maintenance standards. NETA acceptance testing (ATS) applies for the Day 1 testing, and NETA maintenance (MTS) applies after initial testing. For example, in some cases, cable maintenance testing utilizes a lower test voltage compared to the acceptance testing voltage since the cable is no longer brand new. The frequency of maintenance testing varies when looking across the many maintenance standards. Some references provide finite intervals and others have modifiers. For instance, instead of a set interval for maintenance, NETA maintenance assigns a multiplication factor depending on the equipment condition and criticality level.

For example, routine maintenance for metal-enclosed busway:

Maintenance Activity	NETA Frequency (Months)
Visual	2
Visual / Mechanical	12
Visual / Mechanical / Electrical	24

Table 1: Normal Maintenance Intervals

That same metal-enclosed busway, in average service condition and used for a high-reliability application:

Maintenance Activity	NETA Frequency (Months)
Visual	1
Visual / Mechanical	6
Visual / Mechanical / Electrical	12

Table 2: Enhanced Maintenance Intervals

NETA offers a frequency of maintenance document free as a PDF download on their website:

https://www.netaworld.org/standards/frequency-maintenance

NFPA 70B also contains similar data, such as Annex L: Maintenance Intervals. NFPA 70B-Chapter 12 and Annex K include information for long-term maintenance intervals.

Depending on the equipment, the methods will vary. In general, maintenance methods consider these types of activities:

- **Visual:** As the name suggests, this is a walk around the equipment and possibly opening control compartments. Issues stand-out: dirty, abnormal sounds, improper alignments, physical damage, moisture, signs of aging, excessive vibration, corrosion, etc. During this step, it is common to extract equipment historical information from meters and HMI screens.
- **Mechanical:** Cleaning, reseating connectors, confirming torque, replacing components, changing filters, checking bolted connections for high-resistance using tools (torque wrench, low-resistance meter, infrared camera), checking physical operation if applicable, updating firmware, inspecting heaters and their controls, lubricating, etc.
- **Electrical:** Confirming insulation resistance and continuity tests, checking the accuracy of instruments, conducting specialty tests like turns-ratio testing for a transformer, contact resistance, etc.

During a scheduled outage, visual inspection of equipment internal components is recommended. However, some organizations utilize energized work permits to perform energized inspections safely. Be cautious, as even opening a control cubicle containing only 120V likely requires wearing PPE, per NFPA 70E.

Maintenance activity involving mechanical and electrical acceptance tests should be completed while the equipment is in a de-energized state. Maintenance testing compares results to the Day 1 acceptance testing baseline data. Maintenance Teams trend this data over time to inform their maintenance program.

Another measured approach to maintenance: electrical equipment peak loading readings are recorded (monthly) for trending purposes. Maintenance programs then gauge the pace of maintenance first against the recommended industry standards and may increase the frequency of maintenance for parts of the electrical distribution that experience the highest demand loading.

As equipment ages, scheduled in-kind equipment replacements or upgrades become necessary. Electrical distribution equipment typically has service life between twenty-five to fifty-years. These types of factors are used to gauge equipment replacement cycles:

- Well maintained
- Replacement parts/service: remain available
- Environment: suitable for the equipment listing (NEMA/ANSI)
- Loading: high electrical demand or large load profile swings increase wear

Electrical equipment, such as UPS systems, that have active power electronics are usually replaced within fifteen to thirty-years – typically for the sake of newer more energy efficient technologies. Energy storage systems have refresh cycles in the three to twenty-year range.

Arc Flash Safety

Arc flash studies are fundamental in an electrical equipment maintenance program, as they help quantify the potential hazards of working on energized equipment. Note the risk of an arc flash occurring is a combination of the hazard and likelihood of the event. Consider filling out a job hazard analysis before starting electrical work (NFPA 70E, Annex F). See the Electrical Systems Safety section for further information.

The study is typically performed due to a Design Team specified Power System Study requirement, which may also include other focuses, such as breaker coordination, load flow, etc. According to NFPA 70E, Owners should review building arc flash studies every five-years. The standard is not asking Owners to needlessly re-perform arc flash studies, but definitely after any major system change. Even if the building electrical system has not gone through updates or major renovations, arc flash standards are continually changing to refine the calculation methods. Example standards: IEEE 1584, NFPA 70E, and NESC. The same arc flash study conducted five-years ago, if repeated today with the most current standards may yield different results. For example, newer updates consider the physical size of electrical enclosures:

https://www.easypower.com/ieee-1584-2018

Revisiting the arc flash study is a safety-forward approach. If the Owner is willing to pay for a software license, they might consider seeking training on the study software from the Engineer, Contractor, or software provider. At the project completion, a digital copy of the final power system study model should be turned over to the Owner; then, in-house Maintenance Staff may maintain the model with regular updates and check specific arc flash scenarios. Otherwise, the Owner has a record of the model to share with consultants as needed.

Seasonal Testing

If contracted, the Commissioning Team performs a warranty site visit, which is typically a requirement for a formal commissioning process. The visit is an opportunity to discuss the system performance and any issues the Owner Team has found since the building turnover. The Commissioning Agent may also have the chance to take measurements of the system with a different outdoor environmental condition and higher occupant or electrical process loading. Seasonal testing is usually at the ten-month mark to review the systems before the Installing Contractor's standard twelve-month warranty expires. For mission-critical facilities, this warranty visit may just review monitoring rather than functional testing to avoid downtime. The Commissioning Team should provide meeting minutes and a brief report of any testing or measurements from this seasonal visit. In some cases, the results of seasonal testing may require updates to the System Manual documentation. Over time, the Owner may also evolve the System Manual to be a site-specific CFR.

Suppose the building's use pattern fluctuates or has a rapid electrical load increases. In that case, the Owner Team might consider having some level of re-commissioning of the building every three to five-years.

Chapter 4: Safety

Advocate for Site Safety

Safety seems obvious: be safe. However, Electrical Commissioning Team members serve a unique role. They lead an effort to confirm a new or retrofitted electrical system. The very nature of acceptance testing and commissioning is an abnormal condition. The process involves temporary installations potentially subject to these kinds of requirements:

- Project specifications
- Manufacturer instructions
- Company safety practices
- Site-specific requirements
- Utility service requirements
- Applicable local requirements
- National Electrical Safety Code
- OSHA 29 CFR-1910 and 29 CFR-1926
- NFPA 70B-Chapter 7: Personnel Safety
- IEEE 902-Chapter 10: Safe Electrical Work Practices
- NEC; in particular, Article 590: Temporary Installations
- NFPA 70E: Standard for Electrical Safety in the Workplace
- IEEE 1584: Guide for Performing Arc-Flash Hazard Calculations

Figure 8: Lock and Tag

Manufacturer Representatives and on-site Contractors look to the Commissioning Team for testing steps. Being in this role is one of a higher responsibility to advocate for site safety. For that reason, before diving into the technical aspects of electrical commissioning in the upcoming chapters, this chapter highlights the importance of safety. We first look at general site safety and then electrical specific safety suggestions.

Please note there is an entire industry of individuals focused on construction and electrical safety. Those leaders, along with the latest industry codes and standards, should be your deciding references. The discussion in this guide overviews best practices based on project experience.

If you have participated in the commissioning of any building system, you know there is a period of downtime before the sub-systems align for a complete installation. When systems are finally ready, everyone wants to start immediately. The faster they start testing, assuming no issues, the sooner everyone gets to start the countdown clock for the extended burn-in portion of the test (if applicable). People may be in a hurry to wrap up their day at a reasonable hour. The Commissioning Agent should be the person to pause for longer pre-checks and mock test runs. Starting the test thirty-minutes faster will not be worth it if a second attempt is required, equipment faults due to an oversight, or people get hurt.

For example, except for mechanical verifications, the start-up activity will be the first time an Automatic Transfer Switch (ATS) is checked for closed-transition functionality (if included), where the two incoming power sources electrically combine for a temporary period. If the install is incorrect, there could be injuries and/or damage to equipment.

There is not just one safety officer on a job site. All of us are safety officers looking out for each other. For example, as commissioning progresses, the workdays become longer. Everyone associated with the testing process needs to have lunch/dinner and sleep before the next day of testing starts. Working until 11PM and starting at 6AM to maintain the schedule is not sustainable. Having food delivered to the job site is a strategy to maintain progress by not letting the on-site team perform work for extended periods while distracted by hunger. See NFPA 70E, Annex Q: Human Performance and Workplace Electrical Safety.

It takes leadership to delay activities for a group of people to ensure safety. As they say, safety does not happen by accident. Take the time and precautions to prioritize safety. Taking safety seriously is contagious. Others will follow your example because they know it is the responsible approach. As a former mentor once emphasized, "respect the voltage."

General Site Safety

Every job site has site safety requirements that must be followed. Below are good general site safety recommendations for working on construction sites (non-inclusive list):

- Have access to basic tools
    - Flashlight
    - Multi-meter
    - Tape measure
    - Insulated tools
    - Live-voltage indicator
    - Extendable inspection mirror
    - Lock-out tag and key-based hasp lock
- Drive the speed limit on the job site
- Use fiberglass ladders for electrical work
- Use caution with multi-wire branch circuits
- Obey caution/warning tape and safety cones
- Use caution when maneuvering removable floor tiles
- Have a fire extinguisher rated for electrical fires available

- Dress for the conditions
  - Work pants without rips/tears
  - High-visibility long-sleeved shirt
  - Remove dangling jewelry or earrings
  - Wear Personal Protective Equipment (PPE)
    - OSHA approved eye protection and hard hat
    - Personal face covering or listed face mask for some construction activities
    - Flame/arc-resistant clothing, as suggested by NFPA 70E. A full arc flash suit may be required based on activity and risk level.
    - Ear protection: inner-ear and also over-ear if near loud equipment for long periods, such as extended burn-in testing
    - Gloves selected for the appropriate voltage class. Also, gloves to meet the application – running generators are hot to the touch.
  - Closed-toe sturdy shoes/boots with good tread (steel toe may be required)
- Use GFCI circuits for cord-and-plug tools and test equipment
- Tie-off for fall protection at edges of roofs and when climbing over six feet high
- When cell phone coverage may not be available, use handheld radios for coordinating with team members on the job site. Make sure the person repeats back the instructions for final confirmation before proceeding.
  - See NFPA 70E, Annex Q, for the three-way communication process – this is crucial for site safety when operators are communicating between remote buildings.

Electrical Systems Safety

Why are you performing live-work? The answer to this question is usually: "to avoid system downtime." Even in less redundant mission-critical applications, the Owner may have an option to fail-over their computing operations, via the cloud, to a remote site to maintain a no energized work policy. If at all possible, avoid energized electrical work. Today, many companies mandate no energized work as a company policy.

A no-live work policy is ideal, but live-work may still need to occur for new installations. The Commissioning and Contractor Teams must review each instance on a case-by-case basis. Everyone who will be exposed to energized systems should wear appropriate PPE. Reference NFPA 70E, Annex H, and M for guidance on PPE selection. Also, be sure to reference the most current version of NFPA 70E.

Visit NFPA 70E, Annex J, for an example energized work permit. The energized work permit has requirements for Management Staff to sign the permit. This enforcement holds the employer responsible for electrical safety-related to employees; although, it does not reduce the employee's responsibility for electrical safety. Live work should go through proper safety checks, such as live-dead-live testing, which ensures the meter's proper operation. Two further references on this subject (PDF downloads):

https://www.testequipmentdepot.com/application-notes/pdf/safety/electrical-testing-safety-preparing-for-absence-of-voltage-testing_an.PDF

https://lclawards.co.uk/media/70657/select-guidance-for-safe-isolation.pdf

Extreme caution is needed for applications such as large campus utility building or hyper-scale data centers, as these applications typically have modular aspects. De-energizing equipment that all looks similar increases the possibility of human error.

Typically, Infrared (IR) scanning does not occur where incident energy is over 40 cal/cm^2. The design may have included IR windows in the equipment to avoid having workers exposed to the energized interior. For equipment with calculated incident energy over 40 cal/cm^2 (without IR windows), while the equipment is de-energized and locked-out, check the torque marks of all connections or use a low-resistance ohmmeter at those locations.

For calculated arc flash incident energy levels to apply, all protective device settings for the entire system must be installed. Keep in-mind, for secondary services there may not be design techniques for lowering the main's line-side arc flash. A secondary service is when the utility owns the transformer, and the secondary conductors are unprotected, except at the primary side of the service transformer. Depending on the available fault contributions, since the Design Engineer does not have the authority to design the utility's primary equipment, the arc flash levels on the line-side of the service entrance equipment may be excessive. As long as the equipment condition is good and enclosed, as required by the product listing, according to NFPA 70E, the normal open/close circuit breakers' operation does not require PPE. Consider utilizing remote operators, cord-and-pendant open/close pushbuttons, or remote racking robots are further risk mitigation techniques. See the Arc Flash Safety section in the maintenance discussion for further information.

When performing procedures, as a means of change-control, consider a second level of on-call support. It works like this: on-site technicians have an assigned on-call technical person. This individual is knowledgeable regarding the on-site planned testing or maintenance activity. Upon experiencing an unanticipated event (not per the planned script), the on-site team calls the remote staff person to explain which step in the process caused an issue. The on-site team proposes a process modification. The on-call staff member considers the problem telephonically and agrees with the on-site team's procedure change or offers other recommendations. Processes like this are responsible efforts to enhance safety; they also delay testing until the group agrees on the written test procedure.

Reference standards like NECA, NFPA 70B, and Manufacturer literature for safety insights with specific applications. For example, fuel storage, energy storage, remote-operated systems, and rotating machines.

Recommendations for Building Owners:

- Post record electrical one-line diagrams, laminated, in the electrical rooms
- Ask the Design Team to specify safety right into the design by having the Contractor (where it makes sense) provide PPE stations with ear and eye protection
- In non-public spaces, consider floor taping of the required clearance around electrical equipment to discourage storage in the area
- Consider having the project specify the Lock-Out-Tag-Out (LOTO) boxes and include:
    - Push-button locks
    - Portable breaker hasps
    - Group lock-out devices
    - Extension cord plug locks
    - Breaker extension handles
    - Cable for multiple lock-outs via a single lock
    - Locks and tags: Keys have numbers and letters on them. The locks should have corresponding labels to save from trying multiple keys to open a single lock.

Employees who perform electrical work should have safety training; this is in addition to OSHA mandated training for fall protection, eye wash, or otherwise. For example, posted incident energy at a piece of electrical equipment requires technical knowledge to determine the application's proper PPE. Providing training helps employees protect themselves. Even for consultants who only walk job sites for field observation reports (no hands-on installation activities), OSHA Electrical Safety training as an unqualified worker may be appropriate.

Chapter 5: Power Characteristics

Commissioning electrical building systems involves a detailed understanding of power characteristics. The Electrical Commissioning Agent uses their subject matter expertise to communicate testing needs. For example, NFPA 110 site acceptance testing may be required to occur at the generator's nameplate rating (reactive-resistive load banks). See the Load Requirements section for further information.

By no means is the Commissioning Agent *the* only expert. Today's complex electrical systems require an entire industry of support: Contractor, Manufacturer, Design Engineer, Code Official, etc. Even the Author of this guide occasionally needs to revisit these and other topics to reinforce concepts. It is through the fundamentals that electrical systems may be properly confirmed for their applications.

This chapter includes a condensed refresher on electrical theory, intending to aid the Commissioning Agent in establishing the electrical test/monitoring conditions and also reviewing the results. We limit discussion to core insights in electrical system analysis. Despite efforts to be thorough, some topics are not included due to the vast amount of information available concerning electrical theory. Further study is recommended, including reading test equipment manuals, to gain a complete understanding.

The refresher is most helpful to individuals already familiar with electrical systems and responsible for test equipment setup, configuration, or interpretation of test results. If your background does not include electrical theory, after reading, this guidc hopes you know "just enough to be dangerous" – as they say.

A secondary goal of this chapter is to act as a quick reference by grouping concepts. The topics are purposely brief, as there are many other industry references, in addition to experience, needed to understand and appropriately apply methods in electrical analysis. The brevity, hopefully, helps the technical subjects be more consumable as well.

Direct-Current (DC) Circuits

Ohm's Law is shown for the Direct Current (DC) circuit (Figure 9). The unit for voltage is Volts (V), current is Amperes (A), and resistance is Ohm's (Ω). Examples of resistive loads: hairdryer, incandescent lamp, toaster, etc. The higher the voltage, the lower the load demand current will be, as the load's resistance is a property of the device that does not change (constant).

$$\begin{aligned}\text{Voltage}(V) &= \text{Current}(I)\text{Resistance}(R)\\ &= IR\\ &= 2A(5\Omega)\\ &= 10VDC\end{aligned}$$

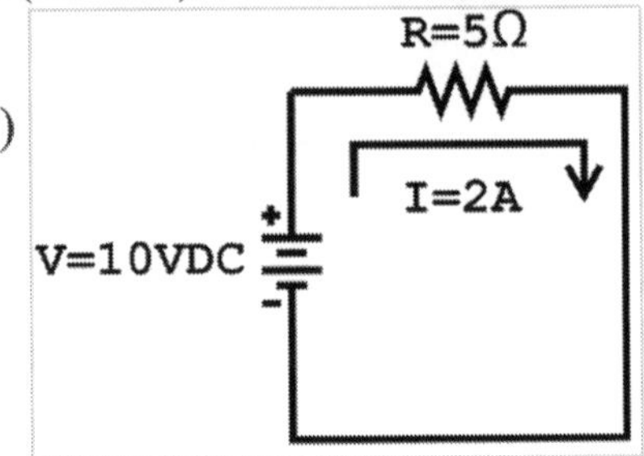

Figure 9: DC Circuit

<u>Instantaneous Power</u>

Power for the Figure 9 DC circuit may be calculated in multiple ways, as seen in the formula wheel. Instantaneous power is measured in Watts (W).

$$\begin{aligned}P_{Inst} &= VI\\ &= 10V(2A)\\ &= 20W\end{aligned}$$

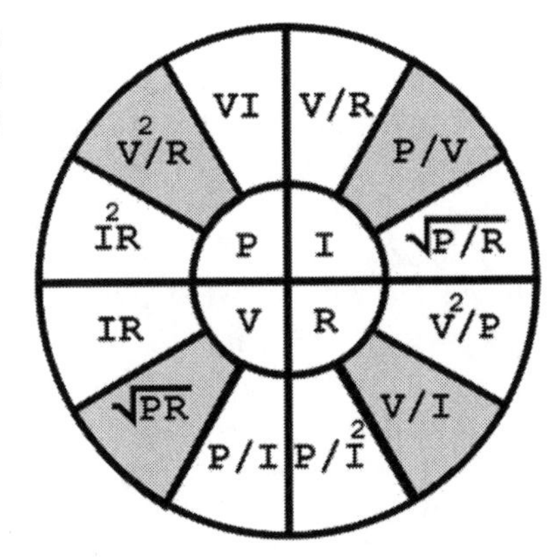

Figure 10: Formula Wheel

<u>Utility Bill</u>

The instantaneous power (W) value is then multiplied by hours used (how long power was drawn for) to compute the utility kilowatt-hour (kWh) charge. For example, the DC circuit example is 20W. If the circuit were powered for four-hours, the total is 0.08kWh. If the utility cost is $0.14/kWh, after approximately fifteen-days, the total cost to the customer (absent other fees) would be $1.00.

$$\begin{aligned}P_{Utility} &= P_{Inst}Time_{Hour}\\ &= \frac{20W(4hours)}{1000}\\ &= 0.08kWh\end{aligned}$$

In addition to energy consumption charges, non-residential buildings are often subject to an electric utility demand rate charge; this is the maximum amount of power (kW) drawn over a fifteen-minute interval and multiplied by a more expensive rate. The non-residential utility bill itemizes the energy consumption and demand charges separately.

System Efficiency

All system components have inefficiencies – not to be mistaken with power factor. System losses are usually in the form of heat dissipated to air. For example, building energy codes might limit the design of feeders to a 2% maximum voltage drop; this is due to the losses in the circuit conductors (since they are material components in the circuit with electrical characteristics).

AC Circuits

Building electrical systems are primarily Alternating Current (AC). Then, local to each piece of utilization equipment, the power is converted from AC-to-DC if needed. For example, a personal computer may utilize 12VDC. The computer's Power Supply Unit (PSU) converts the supplied 120VAC, 60Hz sinusoidal source, from AC-to-DC. An across-the-line motor, for example, would use AC without converting to DC. The primary difference between AC and DC: AC circuits alternate at a periodic rate, and DC is a constant value (not alternating). Calculations for AC circuits may be performed in radians or degrees.

$$\alpha_{(Degrees)} = \frac{\alpha_{(Radians)} 180^{o}}{\pi}$$

$$\alpha_{(Radians)} = \frac{\alpha_{(Degrees)} \pi}{180^{o}}$$

Figure 11 shows an AC waveform as cosine function. The graphs also include a positive constant DC (not alternating) source, shown as a dashed line.

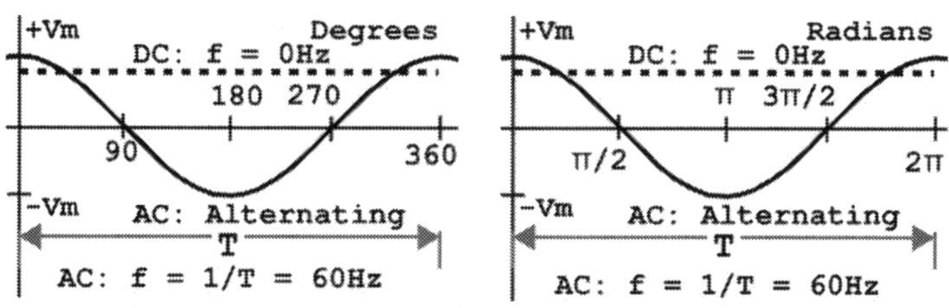

Figure 11: Cosine Waveform (Degrees and Radians)

AC sources alternate the direction of voltage and current in the circuit conductors: positive, negative, positive, negative, repeating. Figure 11 shows one cycle of a cosine function that then repeats itself over and over again. The reason for electrons going back and forth, sixty times each second, is because of the source power. For example, a synchronous generator source rotates a permanent magnet inside an alternator. The magnet's spinning motion induces a voltage in the alternator's conductors that peak negative and positive magnitudes.

AC circuits are represented by a series function – either in sine or cosine format:

$$v = V_m \cos(\omega t)$$

, where:

ω	= the angular frequency
	= 2πf
	= 2π60
	= 376
V_m	= the peak waveform amplitude
t	= time in seconds

RMS vs Average vs Max

For a DC circuit, to compute instantaneous power, we multiply constants together. In an AC circuit, the voltage and current waveforms keep alternating, so the average (mean) value of the waveform is used to calculate instantaneous power characteristics; however, there is one problem. Consider the periodic voltage in Figure 12. Upon integrating the voltage waveform (adding areas under the curves, over time), due to the symmetry, the calculation proves what is visually seen: zero, or very close to zero for the AC waveform average value.

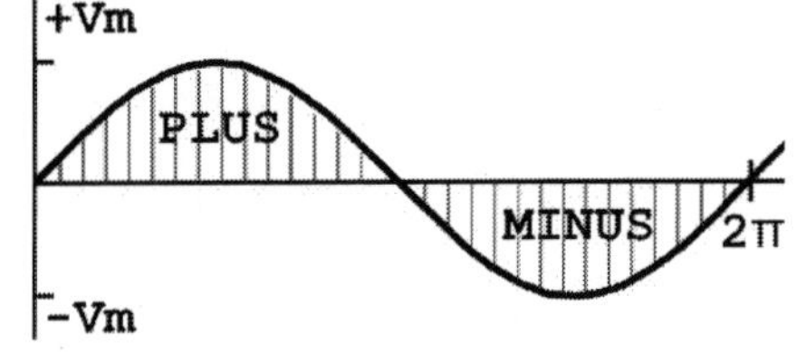

Figure 12: AC Waveform

One may have considered simply ignoring the minus portion so that the average would be the plus portion of the waveform; however, the half-wave cycle average does not equate to the same instantaneous DC value.

$$V_{Av} = V_m 0.637$$

The fix for this is through a mathematical operation known as Root-Mean-Squared (RMS):

$$V_{RMS} = \sqrt{\frac{1}{T}\int_0^T V_m^2 \cos^2(\omega t)dt}$$

, where:

ω	= the angular frequency
V_m	= the peak waveform amplitude
t	= time in seconds
T	= period

The RMS function essentially drags the entire waveform up on the graph (to be all positive, but still alternating). The result is an RMS value that is the AC equivalent to a DC instantaneous voltage or current measurement. By integrating one period, the conversion from RMS to max/peak voltage is:

$$V_{RMS} = \frac{V_m}{\sqrt{2}}$$
$$= V_m 0.7071$$
$$V_m = V_{RMS}\sqrt{2}$$
$$= V_{RMS} 1.4142$$

The RMS value of the waveform is 1.11 times the average value (ratio of 0.7071 / 0.637).

In working with Power Quality Meters (PQMs), the difference in max/peak versus RMS value is readily apparent. For example, instead of 480V, the waveform on a PQM may alternate between +678V and -678V (1.4142 * 480VRMS). Meter settings might be adjustable to view either RMS or max/peak values. Although the context in this discussion has been voltage, RMS calculations for circuit current are the same.

In AC power systems, the RMS values are utilized for power calculations; this updates our DC equations to simply multiply RMS voltage and current values. For example, standardized voltages such as 208V, 615V, 13.8kV, etc.

Utilizing RMS values results in calculations similar to DC. To determine power in AC circuits:

$$P_{Inst} = V_{RMS} I_{RMS}$$

Although, due to Power Factor (PF), this equation does not quite describe the waveform (Figure 13).

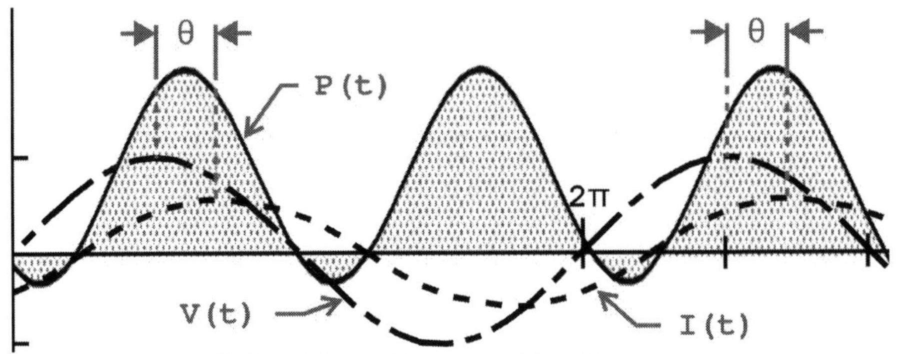

Figure 13: Phase Shifted Voltage-Current and Resulting Power

<u>Power Factor</u>

Reviewing Figure 13 carefully, notice the voltage and current peaks are not aligned (designated by the angle θ). This phase shift means the simple multiplication of voltage and current (as done with DC) does not yield the real power supplied to the load (shown shaded). Instead, the waveform offsets are described:

$$v(t) = V_m \cos(\omega t)$$
$$i(t) = I_m \cos(\omega t - \theta)$$

The offset voltage and current waveforms are still 60Hz (repeating sixty times each second), but the system components are not purely resistive. Building systems include a multitude of loads and sources that cause shifts in the waveforms. The phase shift equates to a system PF. The total single-phase AC power (P), considering PF, is shown below. Here RMS is noted, but it is assumed otherwise:

$$PF = \cos(\theta)$$
$$P = V_{RMS} I_{RMS} PF$$
$$= V_{RMS} I_{RMS} \cos(\theta)$$

Power factor may be leading (capacitive), lagging (inductive), or unity (purely resistive). Most buildings have a lagging power factor, which is ideal to avoid reverse power. Low power factor may need to be addressed to prevent overloading system components and/or satisfy utility requirements. For a leading PF, the current waveform leads the voltage waveform. For a lagging PF, the current waveform lags the voltage waveform. Unity PF (1.0) means the voltage and current waveforms are in-phase.

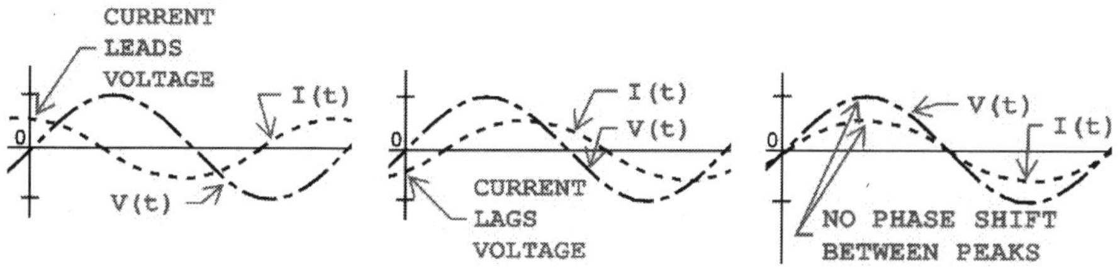

Figure 14: Leading, Lagging, and Unity Voltage-Current

For example, a 0.98 PF lagging is an excellent PF (small voltage-current waveform phase shift), 0.75 PF lagging is rather bad (large phase shift), and 0.95 leading is likely not acceptable (small phase shift, reverse power).

It is accurate to say PF is a measure of how effectively the loads are utilizing source power. Notice power factor is not an inefficiency but instead withheld energy in the AC power system. The power is still useful, but it is merely "stuck" in the electrical system. Utilities have to provide more power to customers with low lagging power factor, even though the customer is not using the power to do real work.

Phase shifted voltage and current waveforms are due to reactive components: capacitors and inductors. For example, see the series RLC circuit in Figure 15. In building systems, many electrical components contain capacitors and inductors. For example, non-linear loads (to be discussed in the Harmonics section). Even the circuit conductors themselves, when used for AC, have reactive characteristics. For less than fifty-mile lengths, circuit conductors are primarily resistive and inductive – for longer lengths, capacitance is also considered. Capacitors resist change in voltage, and inductors resist change in current flow – it is this energy storage in the electrical system that shifts the voltage and current waveforms to create a phase shift that impacts system power factor. In the True Power Factor section we will discuss how harmonics impact PF.

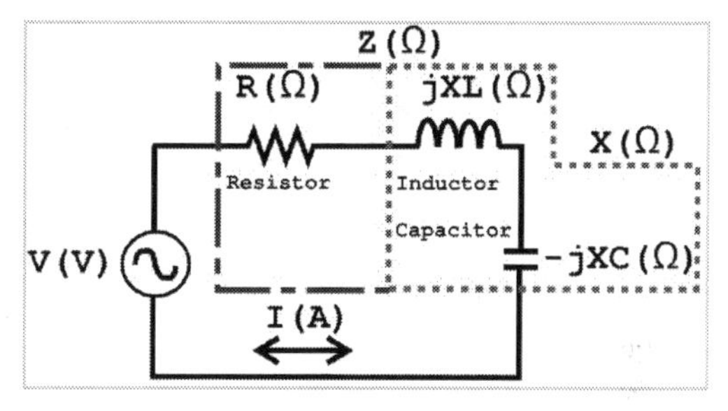

Figure 15: Impedance Circuit

Due to the reactive components, instead of purely resistive, AC circuits have an impedance (Z) defined by:

$$Z = \sqrt{R^2 + X^2}$$

, where:

Z	= impedance (Ω)
R	= resistance (Ω)
X	= reactance (X_L-X_C), or inductance minus capacitance (Ω)
	X_L (Inductance) = $2\pi fL$
	X_C (Capacitance) = $1/(2\pi fC)$

You may notice the impedance formula matches the mathematics expression known as the *Pythagorean Theorem*, which leads us to the discussion of right triangles in the next section.

Apparent Power and Impedance Triangles

In AC electrical systems, the power and impedance triangles are vector representations of the power characteristics. The phasor form represents the frequency domain and shows vectors with both magnitude and direction; this is commonly seen on PQMs and used by the Electrical Commissioning Agent to review power quality relationships.

In Figure 16, the Q and X vectors (pointing up on the page) represent a leading power factor (capacitive). If these triangles had X and Q vectors pointing down (negative angle), that would represent a lagging power factor (inductive). For a unity power factor, Z = R and S = P (phase angle equals zero). The P and R arrow directions left/right also signify providing or receiving power.

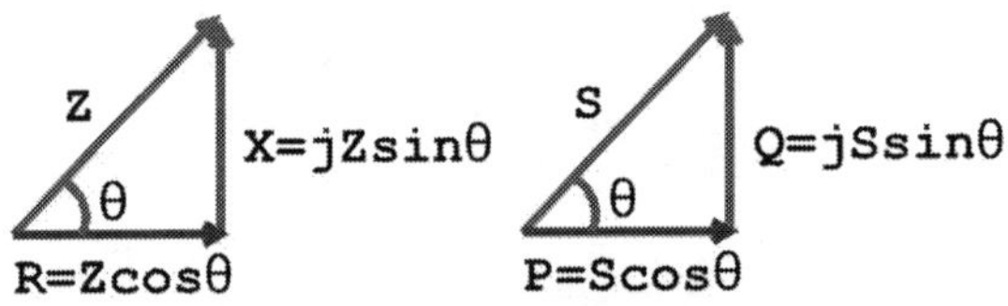

Figure 16: Power and Impedance Triangles

, where θ	= phase angle
R	= resistance (Ω)
X	= reactance (X_L-X_C), or inductance minus capacitance (Ω)
	X_L = $2\pi fL$
	X_C = $1/2\pi fC$
Z	= impedance (Ω)
P	= real power (W) (powering loads rather than system charge)
Q	= vars (energy storage)

Although the use of a positive or negative sign for PF may be tempting, the industry avoids signs to reduce confusion. Instead, leading or lagging after a given PF value is the notation. The up/down vector notation may depend on where (in the world) you are performing analysis – the convention may be reversed compared to what we have noted in this guide. Using leading and lagging terminology also helps avoid confusion about the arrow up/down direction.

With the phasor representations (vector components), the voltage and current waveform phase shifts are seen by the triangle's angle θ. There are several equations for PF. Some of the most common:

$$PF = \frac{\mathrm{Real}Power(P)}{ApparentPower(S)}$$

$$PF = \cos\theta$$

$$\cos^{-1} PF = \theta$$

, where θ = phase angle

Recall helpful mathematics operators (similar for the impedance triangle):

$$\cos\theta = \frac{P}{S} = \frac{\mathrm{Re}\,alPower}{ApparentPower} = \frac{adjacent}{hypotenuse}$$

$$\sin\theta = \frac{Q}{S} = \frac{\mathrm{Re}\,activePower}{ApparentPower} = \frac{opposite}{hypotenuse}$$

$$\tan\theta = \frac{Q}{P} = \frac{\mathrm{Re}\,activePower}{\mathrm{Re}\,alPower} = \frac{opposite}{adjacent}$$

The Apparent Power (S) is both the load power and the system energy storage (Q) caused by the voltage and current waveform phase shift (reactive power, units VAR). It is the "apparent" power (units VA) that accurately describes the total source power required; this is the power triangle's hypotenuse.

$$S = V_{RMS} I_{RMS}$$

$$S = \sqrt{P^2 + Q^2}$$

Frequency Domain

The sine waves shown in previous sections are in the time domain, such that the voltage and current functions, V(t) and I(t), are graphed against time. PQMs display waveforms against time (sinusoidal waves) and in phasor form (power triangle). The frequency domain allows graphing of the waveforms against frequency (Hz) instead of time (t).

Using a mathematical operation called a Fourier Transform determines the signal's frequency components. Understanding the frequency components is particularly useful when waveforms include harmonics (multiples of the fundamental 60Hz frequency) due to non-linear loads. Non-linear loads will be discussed further in the Harmonics section.

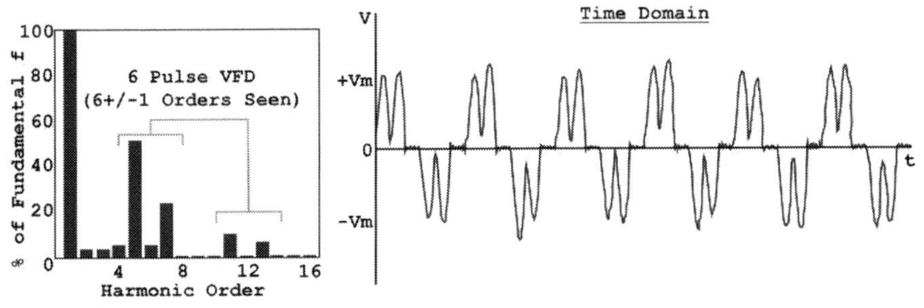

Figure 17: Harmonics Time and Frequency Domain

The building systems being discussed in this guide have a constant 60Hz source. If only this fundamental frequency exists in the AC circuit (no harmonics), a Fourier Transformation would only show one amplitude for the 60Hz (Order 1; the tallest line on the Figure 17 bar graph) – a pure sine wave without disturbances.

Phasors are useful to see the phase relationships between voltage and current quickly. For example, if one phase has a frequency variation in a three-phase system, then the associated phasor shows other than 120 degrees between phases. The frequency domain makes calculations less laborious than they are in the time domain, which requires more detailed mathematics via polar or rectangular forms (complex/imaginary numbers) – outlined in the next section.

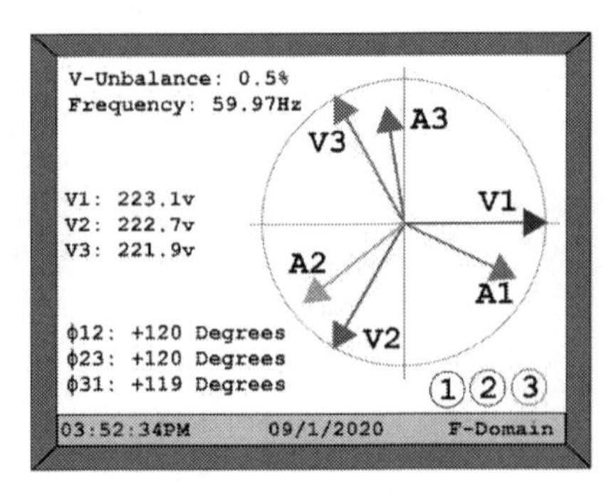

Figure 18: PQM Phasors

<u>Polar and Rectangular Forms</u>

The sinusoidal functions are infinite series functions:

$$\sin x \quad = x - \frac{x^3}{2} + \frac{x^5}{2} - \frac{x^7}{2} + \frac{x^9}{2} - \ldots$$

$$\cos x \quad = 1 - \frac{x^2}{2!} + \frac{x^4}{4!} - \frac{x^6}{6!} + \frac{x^8}{8!} - \ldots$$

Sine and cosine functions are very similar to Euler's number raised to 'x' for the infinite series (base of the natural logarithm, *e*):

$$e^x \quad = 1 + x + \frac{x^2}{2!} + \frac{x^3}{3!} + \frac{x^4}{4!} + \ldots$$

Adding the equations together and including the phase angle in place of 'x' provides the beneficial Euler's formula.

$$e^{j\theta} \quad = \cos\theta + j\sin\theta$$

As seen in the impedance triangle, in Euler's formula, cos θ is the real component ('R'), and sin θ is the reactive component ('X'); this is known as the rectangular form.

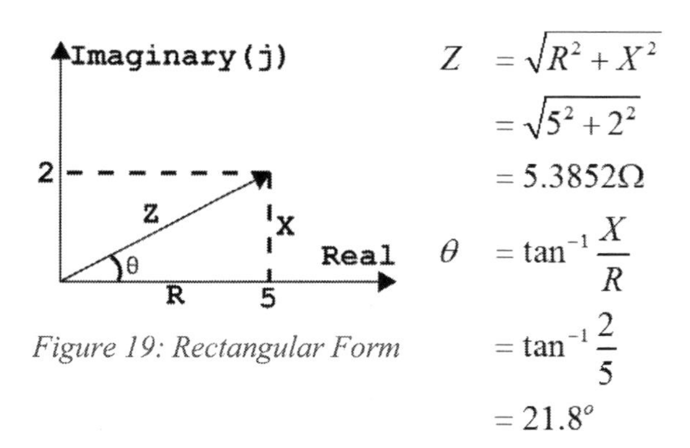

Figure 19: Rectangular Form

$$\begin{aligned} Z &= \sqrt{R^2 + X^2} \\ &= \sqrt{5^2 + 2^2} \\ &= 5.3852\Omega \\ \theta &= \tan^{-1}\frac{X}{R} \\ &= \tan^{-1}\frac{2}{5} \\ &= 21.8^o \end{aligned}$$

The polar form equivalent describes the impedance vector ('Z') with a magnitude and phase angle:

$$\begin{aligned} Z &= |Z|\, e^{j\theta} \\ &= 5.39@21.8^o \end{aligned}$$

The 'j' in Euler's formula is introduced as an imaginary number. Imaginary numbers are also known as complex numbers. Some argue these numbers should be called lateral numbers to avoid the stigma that 'j' is "imaginary" and not useful. In essence, complex numbers allow solutions to complex problems. The imaginary mathematical number 'i' = sqrt(-1) is the same used in power calculations, except Electrical Engineers uses 'j' to avoid confusion with ampacity (i).

Harmonics

Although perfect sine waves are the goal, non-linear loads commonly demand current from the power source in "gulps" or short pulses. Power electronics have control of the input demand and may not pull current continuously. Instead, non-linear circuits demand current to charge internal reactive components (inductors/capacitors), pause, and then draw current again. The circuit may have semi-conductor components that switch on and off to power loads at a different frequency than the 60Hz source power. For example, a VFD (motor controller) converts the incoming AC-to-DC and recreates a digital sine wave at the VFD output – this mimics a continuous sine wave at varying frequency. Specifically, non-linear loads include impedance elements that change with voltage due to the angular frequency component (ω) in $\cos(\omega t)$.

Linear Load Examples: pump, fan, incandescent lamp (light bulb), and other utilization equipment that uses the natural sine wave as received from the source.

Non-Linear Load Examples: computer power supplies, fluorescent lighting, Variable Frequency Drives (VFDs), electric welders, and other utilization equipment that converts the source sine wave for use at a different frequency or voltage (possibly DC).

Take the example of an incandescent lamp or motor – these linear loads would decrease in brightness or slow their speed if the voltage decreases. Non-linear loads, which utilize electronic components or potentially power electronics, which simply pull more current if the voltage decreases (keeping the same power demand). The response to maintain constant power is what their circuits have been designed to do (power the load effectively); thus, non-linear loads distort the AC sine wave and result in system harmonics.

Waveform distortion is easily seen when looking at the time domain. These waveforms are a combination of the fundamental frequency sine wave (60Hz) and any harmonic waveforms caused by non-linear loads. Harmonic waveforms are percentage multiples of the fundamental frequency. For example, a third-order harmonic would have a frequency of 180Hz (60Hz * 3), fifth-order would have a frequency of 300Hz, and so on. The resulting output waveform contains the fundamental frequency, as well as some percentage of the third and fifth-order harmonics.

In Figure 20, on the left is the 60Hz fundamental frequency and the other waveforms are harmonics caused by system loads. Although this example is an extreme case, the resulting waveform on the right demonstrates how the natural sine wave is distorted. Harmonics can be harmful, potentially causing electrical components to operate outside of their listed ratings. Resulting waveforms are referenced as having even, odd, or a combination of harmonics at different orders.

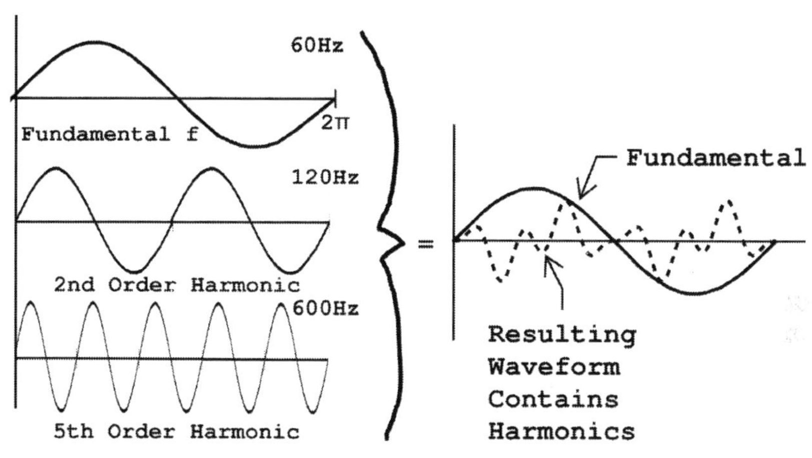

Figure 20: Harmonic Content Waveform

In order to aid in studying harmonics in power systems, it is helpful to utilize a graphing calculator to create test cases of waveform shapes. For example, for a 120VAC branch circuit, type the expression into Desmos (free online option) for the example given in Figure 20:

First click the wrench at the top right to change from radians to degrees

Equation #1: 120*sin(x)

Equation #2: 120*sin(x)*sin(x*2)*sin(x*5)

https://www.desmos.com/calculator

A significant driver of harmonics in years past used to be Switched Mode Power Supplies (SMPSs) for computers; however, today, even most home computer Power Supply Units (PSUs) include Active Power Factor Correction (APFC). This technology adjusts the sine waves to mostly eliminate SMPS system harmonics.

Consider, harmonic circuit content often results in unbalance for three-phase systems; thus, the harmonic current is commonly carried by the neutral conductor. Designers sometimes consider oversizing neutrals in the electrical distribution system and/or specifying K-rated transformers. This additional copper (or aluminum) helps the components carry more harmonic current without exceeding their ratings but does not prevent the harmonics from occurring in the electrical system.

Some harmonics are temporary (not periodic), such as transformer in-rush (Order 2 harmonics, decaying). There are still many other sources contributing to harmonics. See IEEE 519 for further information on Total Harmonic Distortion (THD) limits, which helps protect power systems from the impact of harmonics. The local utility service and Manufacturer equipment manuals define acceptable percentage THD limits also.

See the Unbalanced Systems section for harmonic order as it relates to phase sequence.

True Power Factor

In this chapter so far, we generically refer to "Power Factor (PF)." Specifically, the PF being referred to is the "Displacement PF (DPF)," or fundamental PF. Considering harmonics exist in power systems, the subject of PF needs a little more detail.

By definition, the DPF is the phase shift between voltage and current waveforms. Consider a distorted waveform due to harmonic contributions – this impacts True PF, as non-linear load harmonic current is non-sinusoidal. To distinguish between DPF (the voltage and current phase angle shift), True PF is defined as:

$$PF_{True} = DPF * APF$$

Apparent Power Factor (APF) is the distortion factor caused by non-linear load harmonics. Harmonic circuit content does not result in real work for the loads but instead yields reactive power (stuck in the system, units VAR as seen in the Figure 21 vector diagram).

Using the power triangle, the introduction of harmonic content to a circuit means:

$$S^2 \geq P^2 + Q^2$$

Since the inequality is valid, a new vector 'D' (Distortion) is included in the power triangle. This third dimension ('Z') requires a geometric sum to solve rather than basic math equations.

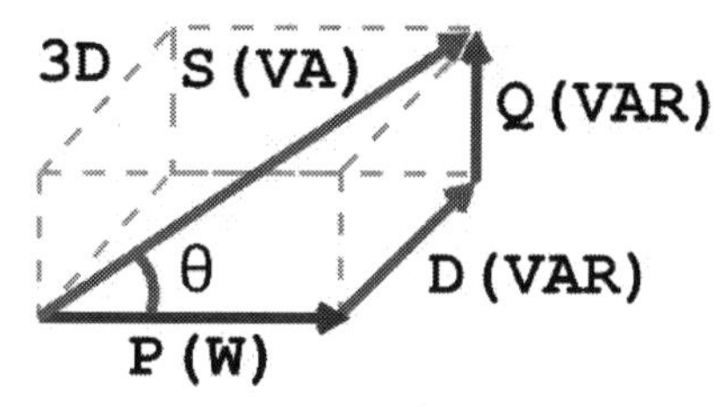

Figure 21: Vectors for True Power Factor

When working with test instruments, both DFP and True PF may be provided. From a design standpoint, standard PF correction techniques alone may not address all problems caused by harmonics and may require active harmonic filtering.

Power Quality

Power quality is an in-depth subject. The industry is fortunate to have several experts in this area. For commissioning, the Electrical Commissioning Agent helps recognize power quality issues and then works with Vendors and the EOR for a solution or a more in-depth study. For example, the power system may experience voltage sags, swells, transients, notches, unbalance, interruptions, frequency deviations, noise, etc. Many of these power quality issues have negative impacts on electrical equipment and overall power stability. Since other sections in this chapter have detailed the ideal AC waveform, we will save the mathematics and qualitative discussion in this section for this most helpful AEMC article titled *"Understanding Power & Power Quality Measurements"* (PDF download):

https://www.aemc.com/userfiles/files/resources/applications/power/understanding_power.pdf

Also consider reviewing NFPA 70B, Chapter 10: Power Quality.

Three-Phase AC Power

In the RMS vs Average vs Max section, voltage and current waveforms are multiplied to show power (designated by the shaded portions in Figure 13). That was single-phase AC power. Including two more phases and separating them by 120 degrees produces more capacity in power systems. A three-phase power system has three times the power of a single-phase system (Figure 22).

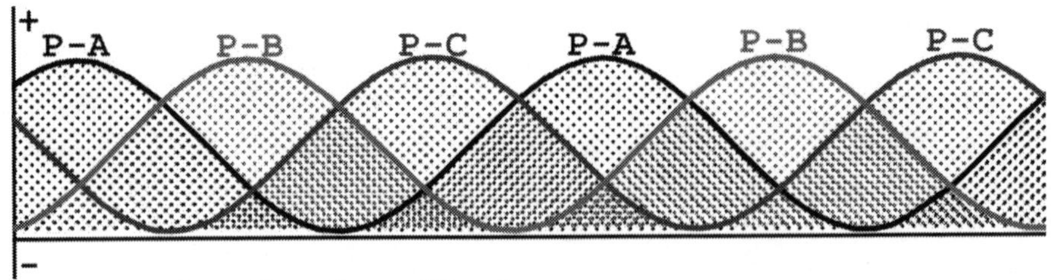

Figure 22: Three-Phase Power Waveform

Why only three phases and not six or more? More than three phases would provide power between the peaks, and each phase would carry less current. Strictly from an economic standpoint, the industry settled on three phases and it has been that way ever since. When three-phase systems expect current values (driving conductor sizes) will be in excess, even after paralleling conductors, the applications simply utilize a higher voltage to lower the current to a more manageable level – still with just the three phases.

For example, try typing the P=VI expressions into online mathematical software Desmos. We will use 208VRMS and 2ARMS for all phases (balanced). Separating the waveforms by 120 degrees offers a similar graph to the Figure 22 (three-phase power).

First click the wrench at the top right to change from radians to degrees

#1	$V_1 = 208\sin(x)$	#4	$I_1 = 2\sin(x)$	#7	$P_1 = V_1 \cdot I_1$
#2	$V_2 = 208\sin(x+120)$	#5	$I_2 = 2\sin(x+120)$	#8	$P_2 = V_2 \cdot I_2$
#3	$V_3 = 208\sin(x+240)$	#6	$I_3 = 2\sin(x+240)$	#9	$P_3 = V_3 \cdot I_3$

https://www.desmos.com/calculator

Figure 22 is for a unity PF (1.0) system, as each phase uses the same phase angle for voltage and current (no phase shift). Note: we describe voltage and current magnitudes with RMS values, but a PQM meter may be set to display peak values (1.4142 * RMS).

Balanced power systems that exclude non-linear loads have linearity, which means a system of related equations may describe them. For example, a transformer's temporary in-rush (when energized by closing a switch or circuit breaker) is non-linear and much more difficult to express mathematically. Electrical Engineers capitalize on linearity to use the concept of Superposition. By this, a balanced three-phase AC power system may be analyzed by reviewing the single-phase power and multiplying by three (due to three phases). Unless there is unbalance in the system, each phase of the three phases does not need to be calculated individually.

Balanced Systems

In the AC Circuits section, due to a possible phase shift in voltage and current waveforms, introduced by reactive loads (inductors and capacitors), a PF is required to compute real power (P). Using Superposition, a three-phase balanced power system may be analyzed by calculating each phase and multiplying by three to obtain the three-phase value. In a balanced system, all three phases carry the same magnitude of current.

$$P(Watts) = 3 * V_{Ph} * I_{Ph} * \cos(\theta)$$
$$S(VA) = 3 * V_{Ph} * I_{Ph}$$

When utilizing the line-to-line values of voltage and current (between phases, rather than line-to-ground or neutral), the three-phase balanced power is determined by multiplying by sqrt(3):

$$P(Watts) = \sqrt{3} * V_L * I_L * \cos(\theta)$$
$$S(VA) = \sqrt{3} * V_L * I_L$$

The load uses Power (P, units W) for real work, and the Apparent Power (S, units VA) is the total power required for the system, which includes the reactive power (Q, units VAR).

Transformation

Due to the properties of AC, the voltage may be transformed higher or lower. DC voltage may be changed via electrical components, but typically results in lower system efficiency (compared to AC) due to the components needed to make the voltage change. In AC systems, transformers are used to step-up or down voltage. Inductors (continuous coils of wire) are connected in Delta and/or Wye configurations. The primary/secondary windings share a common transformer core. Looking at a transformer in-person will appear like there are only three, instead of six (three primary and three secondary), windings.

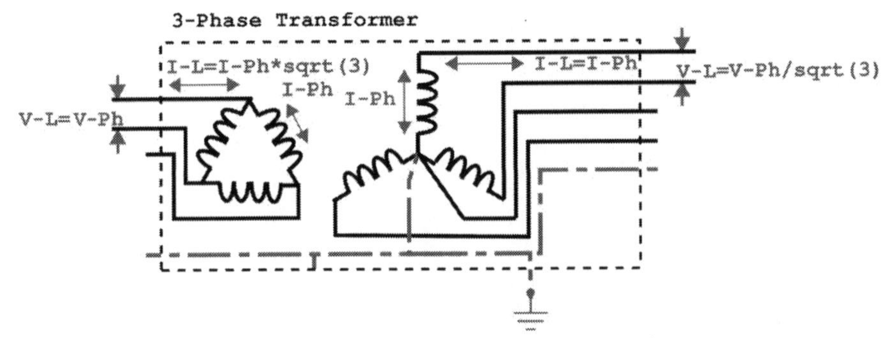

Figure 23: Three-Phase Delta-Wye Transformer

Transformers induce voltage without physical connections between the primary and secondary windings. The ratio of primary/secondary is a multiplier for transformation. For Delta-Wye transformers, as shown in Figure 23, due to three-phases sqrt(3) is used mathematically to convert transformer voltages and currents.

Transformers are manufactured for both single and three-phase AC applications. Delta-Wye transformers are common in building applications; this arrangement inherently introduces a thirty-degree phase shift between the primary and secondary. Transformers are passive, linear components in the power system and usually have an impedance (Z) ranging from 3% to 8% (per unit value) – inefficiency resulting in heat.

Phase Sequence, Rotation, and Arrangement

The utility grid, campus primary and secondary selective systems, branch circuit wiring, and even temporary connections are influenced by electrical phase sequence, rotation, and arrangement. For this reason, as a pre-step to functional performance testing, the commissioning process must take care when confirming these power characteristics.

The industry standard bodies have established the phase arrangement within the equipment. All manufactured equipment bus bars or device terminals are arranged ABC, either left-to-right or top-to-bottom. Depending on where (in the world) the installation is, and possibly the system voltage level, L1L2L3 may be the naming convention rather than ABC. Ideally, equipment interior busbars are clearly labeled by the Manufacturer to provide visual confirmation during maintenance and Infrared (IR) scanning.

In three-phase AC power systems, as seen in the time domain graph (Figure 22), maximum power, per phase, is in a sequence (cyclically, 120 degrees apart): ABC, ACB, BAC, BCA, CAB, or CBA.

Typically, only ABC or ACB phase sequences are noted – keeping A as the first phase since the waveforms repeat. An ABC or ACB phase sequence equates to Clockwise or Counter-Clockwise rotation, abbreviated CW and CCW (anti-CW). For example, a generator powered by a fuel source rotates to induce a voltage, or a motor receives power to physically turn a fan.

In some cases, the phase sequence may not equate to phase rotation. For example, even if a piece of equipment is designed for CW rotation (such as a UPS) and has ABC phase connections, there is an opportunity for wires to become crossed on both the utility and customer side. For small motor applications, an Electrician might perform a "bump test," energizing the motor for a moment to see which direction it starts to spin (CW or CCW). Before doing this test, to prevent damage, the motor should be decoupled from the load. Suppose the rotation is backward based on the spinning direction (such as a fan or pump). In that case, an Electrician physically re-terminates two wires to reverse the rotation, changing the system from ABC to ACB (or vice versa). See Manufacturer's manuals and industry standards, such as NECA and NFPA 70B for further information.

Phase rotation is especially important for the paralleling of sources. Paralleling unsynchronized sources out-of-phase, with a significant frequency difference, or sources with different voltage magnitudes, likely results in electrical and possibly mechanical stress on the system components; that is, with regards to the insulation and physically within the machines. Such conditions may also pose harm to the operator. Before paralleling systems, after a rotation meter check, a voltage check must also be performed. For example, confirm there is nearly zero potential difference (voltage) between Source 1 Phase A and Source 2 Phase A.

As seen in the discussion above, phase sequence and rotation are not easily distinguished in conversations. For this reason, special attention should be given to site labeling. More explicit labeling and identification beyond just the phase sequence may be needed to communicate the system's phase rotation accurately. Even if the rotation is accurately described in writing, it may be interpreted differently. This guide is not aware of a standard that dictates if the rotation is described with respect to the source rotation or the load's resulting torque.

The relaytraining.com website has an excellent summary (with videos) on this subject here:

https://relaytraining.com/understand-determine-phase-rotation/

Unbalanced Systems

One benefit of three-phase power is the multiple voltages that are available. For example, the load may be powered by using two of the three phases. Even though this arrangement uses two phases, the industry refers to this as single-phase. If the source has a grounded (neutral) conductor, single-phase circuits may also be made up of just one phase conductor and the neutral.

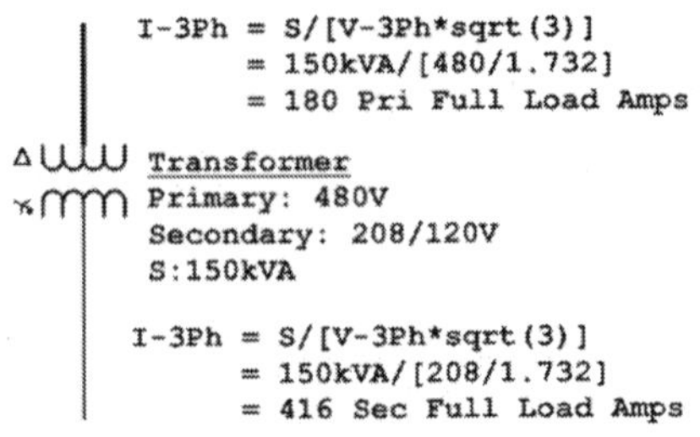

Figure 24: Single-Line Diagram Step-Down Transformer

Example: 150kVA Load	**Voltage / Single or Three-Phase**		
	480 / Three	*480 / Single*	*277 / Single*
Current (A)	180	313	542
Apparent Power (VA)	$=V_{3Phase}I_{3Phase}\sqrt{3}$	$=V_{1Phase}I_{1Phase}$	$=V_{1Phase}I_{1Phase}$
	$=480*180*1.732$	$=480*313$	$=277*542$
	=150kVA	=150kVA	=150kVA

Table 3: 150kVA Transformer Circuiting Examples

Since not all building loads are three-phase and multiple connection options are available, there is potential for unbalance in the power system. Unbalance current flows on the neutral conductor.

In an unbalanced condition, the three phases are not evenly loaded; thus, the principle of Superposition is not applicable (we cannot calculate total power by simply multiplying one phase by a factor of three). Although unbalanced equations may be solved using vector algebra in the frequency domain, this is a time-consuming process, especially if calculations are preliminary and must be repeated for different test cases.

Leaning on the concept of linearity, Electrical Engineers solve unbalanced problems by translating the RMS voltages and currents into Symmetrical Components. The symmetrical component models allow the unbalanced power system to be described with three separate balanced system conditions:

Positive Sequence: Clock-Wise Rotation, 120° Between Phases
Negative Sequence: Counter-Clockwise Rotation, 120° Between Phases
Zero Sequence: No Phase Displacement, Equal Magnitudes

Similar to a Fourier Series mathematics operation that allows changing between the time and frequency domains, the Symmetrical Component operators (not documented in this guide) allow the unbalanced power system to be translated for more straightforward calculations; the reverse operator then converts Symmetrical Components back to the time domain. As an alternative to the Ohmic method, Symmetrical Component analysis may be used for fault calculations; the Per-Unit method (not documented in this guide) is often used with Ohmic and Symmetrical Component approaches.

Electrical phase sequence also allows for classification of harmonic orders:

Positive Sequence: 1, 4, 7, 10, ...
Negative Sequence: 2, 5, 8, 11, ...
Zero Sequence: 3, 6, 9, 12, ...

As demonstrated in this chapter, knowledge of power characteristics is fundamental for commissioning of electrical building systems. By making proper determinations and evaluations of the outputs, the Electrical Commissioning Agent is able to apply these concepts and use test instruments to confirm proper operation of equipment and combined systems.

Chapter 6: Test Equipment Plan

Purpose

A Test Equipment Plan, sometimes referred to as a Load Bank Plan, is a document tool to communicate instruments and load bank rentals needed to commission building systems. This guide is not aware of an industry standard establishing criteria for this kind of plan. Some providers use a Load Bank Plan, and others embed information in specifications or the Commissioning Plan. For smaller projects, this plan may not need to be a formal deliverable.

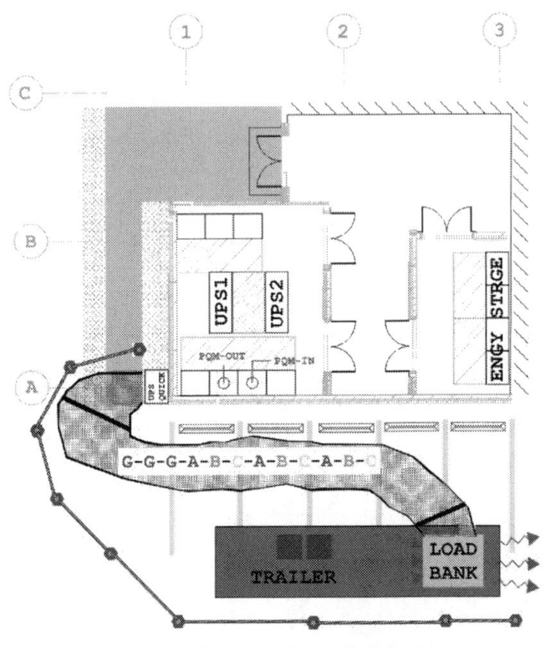

Figure 25: Site Plan Load Bank Setup

Since the Test Equipment Plan is a driver for the planned test activities, it is usually written by the Commissioning Provider. If the Design Team fully specifies the commissioning requirements, the Contractor may also be a suitable party to write the plan. Remember, the document will likely be passed through from the Contractor to a Rental Supplier. The Commissioning Agent should do their best to include relevant details needed to provide a proper quote. For instance, the plan may require a minimum of X feet of load bank cable, and longer lengths are subject to the Contractor's determined temporary routing. A specific load profile may also be required (resistive, inductive, capacitive). It is beneficial for the team to have a meeting to review the plan in detail.

The list below highlights common information communicated by the Test Equipment Plan:

- Load bank cording requirements
- General coordination topics related to testing
- Where and how to connect the test equipment in the electrical system

- Types, ratings, minimum load step resolution, and quantities of load banks and test instruments
- Approximate length of each functional test (amount of time) to allow the affected Contractors to plan accordingly
    - Vendors may have a maximum number of continuous hours their representative can remain on-site. In that case, they may need to coordinate a staff change to complete extended burn-in testing. Or if there is a failure and the test must be restarted, testing could take longer or need to be rescheduled.
- Communication of additional steps, such as connections/disconnections, PQM setup, and enhanced attention to safety that might take more than the scheduled test runtime

Load Banks

Electrical commissioning comes with an interesting challenge: for new buildings, commissioning occurs before occupancy. There may not be a significant electrical demand load without people in the building or processes operating. Testing of electrical systems without load is a necessary step in the process, but there is typically value seen in renting load banks for final equipment confirmations. Load banks come in many types and capacities, but the most basic are air-cooled and made up of heavy-duty resistors or heating elements; they are essentially huge hairdryers with basic circuit control and protection. Reactive load banks are also available where simulation of building loads requires an inductive load, which is a typical load profile for most buildings, or a capacitive load, which can simulate common telecommunications and computer loads.

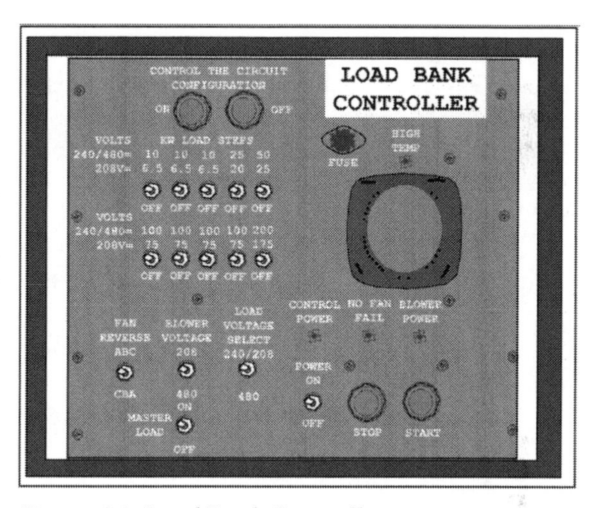

Figure 26: Load Bank Controller

The load bank power demand varies as a human operator selects manual dip switches or changes load steps on an electronic controller. Load banks use internal fans to avoid overheating. Simulating the building load with test equipment is one way to confirm the permanent building systems. Some system response characteristics may be simulated, such as voltage dip upon placing full-load on a standby generator (100% block load). By bringing the load banks inside the facility, the mechanical system can also take advantage of these heat-generating pieces of test equipment to validate cooling design capacities.

<u>Form Factors</u>

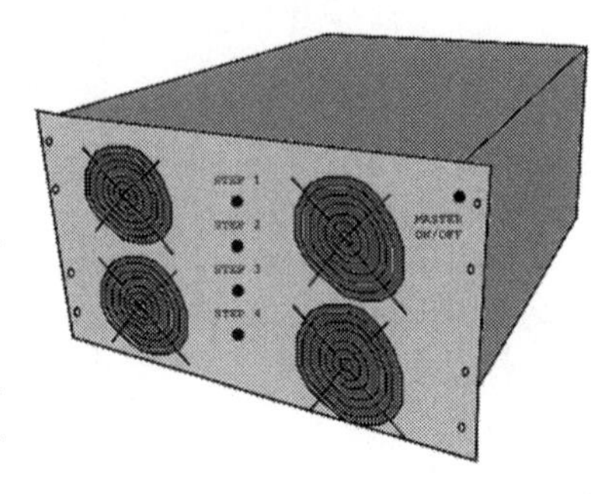

Figure 27: Rack Mount Load Bank

For computer rooms, rack mount load banks are common. Alternatively, small suitcase or rolling style load banks may be spread throughout the computer rooms to simulate even heat distribution. The alternative approach is practical for testing system capacities, but it is a less realistic test than a rack-mount load bank solution. Rack mount solutions have adjustable speed fans. For enclosed aisle containment systems (hot/cold), the fans' contributions may have an impact on the air pressure within the aisle. Active chimneys may play a part in coordination also. Work with the Mechanical Team members to evaluate load bank specifications. Water-cooled rack options are available for laboratory-based user applications but may have limited rental market availability.

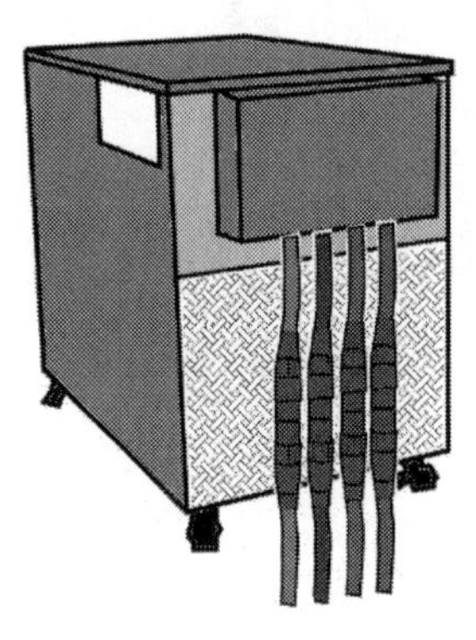

Figure 28: Rolling Load Bank

Rolling load banks (mounted on casters) are usually available up to 500kW. Above this rating, load banks are typically trailer mounted. Load banks with reactive circuit components may also be trailer mounted, as they are heavier than purely resistive options. Trailer mounting is standard for medium-voltage load banks that either operate at the system voltage or require a rental transformer and use of low-voltage (<1000V) load banks. Optionally, some load banks may be configured for air-flow to exit the top.

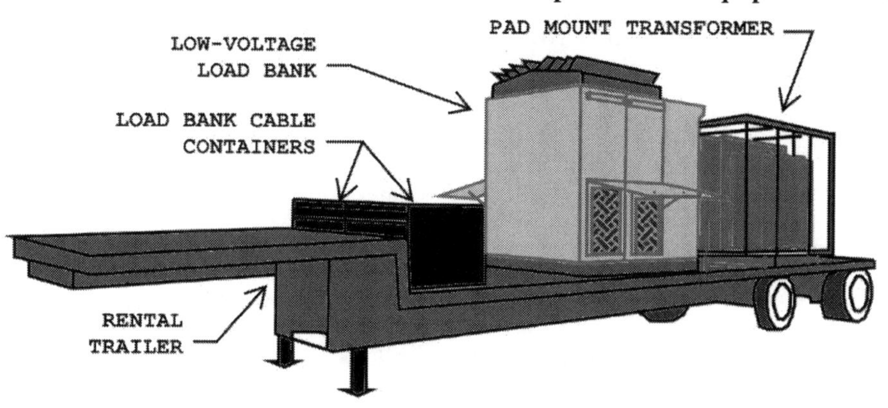

Figure 29: Trailer Mounted Load Bank and Transformer

Types and Features

Although the construction industry only uses a handful of load bank types for building commissioning, here is a brief overview of the most common types:

- **Purely Resistive:** The most common for on-site unity PF (1.0) testing
- **Direct Current (DC):** Smaller resistive load banks, used for battery systems, such as switchgear station power
- **Reactive-Resistive:**
  - Confirms alternator voltage regulation, frequency/voltage transient response, and load sharing. If specified, the Manufacturer or Integrator uses the reactive components for required pre-site delivery (in factory) testing.
  - If the building PF is not close to unity (1.0), the Project Team may determine these are required for site parallel testing to prove and tweak load sharing between generators.
  - Generator PF testing reference (PDF download): https://incal.cummins.com/www/literature/technicalpapers/PT-6004-PowerFactorTests-en.pdf
- **Inductive-Capacitive:**
  - Reactive only circuit components (inductors/capacitors). For instance, generator alternators have very low PF testing requirements that are completed by the Manufacturer before assembling the generator set.
- **Electronic:**
  - Includes report generation and allows the operator to program dynamic load profiles. For example, the operator may simulate discharge characteristics of a particular battery type.

- **Electronic Regenerative:**
  - These contain power electronics that demand power as defined by the user. They do not expel energy like a normal load bank (heat to air or water). Instead, they divert the energy back to the source for powering other building loads. These are a "green" (environmentally friendly) load bank option since the only energy used is the efficiency loss of the internal load bank components.
  - Availability for rental may be limited or non-existent.
- **Water-Cooled:**
  - Instead of using air to cool load bank components, this option provides mechanical systems a way to heat their liquid mediums for chiller, cooling tower, and other similar system components.
  - Unlike air-cooled load banks, water-cooled load banks are reasonably quiet since they do not have large fans running.
  - Closer Test Equipment Plan coordination is needed since these load banks have electrical and mechanical system connection requirements.

<u>Product Examples</u>

The easiest way to specify load banks in the Test Equipment Plan is by using the internet to find a basis of commissioning product that meets the project requirements. Seek out Manufacturers and use their online filter tools to narrow down your search results. Online research helps locate real-world equipment, potentially with reference product specifications. See the Rent Versus Buy: Load Banks section. The following pages provide product examples:

- **Rack Mounted:**
  - ComRent CRLS 11.5kW: https://www.environmental-expert.com/products/comrent-model-crls-11-5-kw-rack-mounted-load-unit-504948
  - Simplex MicroStar: https://www.simplexdirect.com/Product.aspx?ProdID=81
- **Suitcase:**
  - ASCO 2500: https://www.ascopower.com/us/en/product-range-presentation/66158-asco-2500-load-bank/?parent-subcategory-id=89081
  - ComRent K490 (PDF download): https://www.comrent.com/wp-content/uploads/2018/05/K490_Updated.pdf

- **Rolling:**
  - ComRent LPH100 (PDF download): https://www.comrent.com/wp-content/uploads/2018/05/LPH100_Updated.pdf
  - Simplex PowerStar: https://www.simplexdirect.com/Product.aspx?ProdID=39
  - ComRent LPH400 (PDF download): https://www.comrent.com/wp-content/uploads/2018/05/LPH400_updated.pdf
  - Simplex Merlin: https://www.simplexdirect.com/Product.aspx?ProdID=50
  - ASCO 2805: https://www.ascopower.com/us/en/product-range-presentation/66164-asco-2805-load-bank/?parent-subcategory-id=89081
- **Trailer Mounted:**
  - ASCO 4900: https://www.ascopower.com/us/en/product-range-presentation/66181-asco-4900-load-bank/?parent-subcategory-id=89081
  - ComRent K875 (PDF download): https://www.comrent.com/wp-content/uploads/2018/05/K875A_Updated.pdf
  - Simplex Atlas: https://www.simplexdirect.com/Product.aspx?ProdID=55
- **Radiator Mounted:**
  - ASCO 1100: https://www.ascopower.com/us/en/product-range-presentation/66183-asco-1100-load-bank/
- **Medium-Voltage:**
  - MV load bank info: https://www.ascopower.com/us/en/resources/articles/medium-voltage-load-banks.jsp
  - ASCO 9100: https://www.ascopower.com/us/en/product-range-presentation/66210-asco-9100-load-bank/
  - ComRent CR3750 (PDF download): https://www.comrent.com/wp-content/uploads/2018/05/CR3750_Updated.pdf
- **Water-Cooled:**
  - LBW500: https://www.simplexdirect.com/Product.aspx?ProdID=18

This load bank list is only a select sampling of the available products – it is not intended to be an all-inclusive list or as a recommendation for specific Manufacturers. The product examples may only be available for rental or purchase. Listing the example products in the Test Equipment Plan helps the Contractor's Supplier understand the testing intent. Although the exact load bank make/model may be listed, there may be a contractual requirement to list multiple test equipment models and Manufacturers. For instance, government projects may require three Manufacturer options. Many reputable Manufacturers rent/sell load banks, so the specifier should find Equipment Suppliers that align with their project requirements.

Load Requirements

It is helpful if the Test Equipment Plan identifies the minimum load criteria or basis of the commissioning approach. To balance the cost of testing with the level of performance, this should be closely coordinated. For example, there may not be enough rack mounted load banks in the rental market, requiring an alternate load banking approach.

Commonly, the available power at IT racks is higher to allow flexibility in the data center equipment layouts. Coordinate with the Mechanical Team to understand the maximum cabinet power densities. Optionally, if there is a Computational Fluid Dynamics (CFD) model, it would provide even more specific rack loading assumptions for closer to real-world testing. Right-sizing the load banks instead of sizing them based on the maximum electrical capacity available at the individual IT racks helps avoid renting too many load banks and not effectively utilizing them.

During the re-occurring commissioning meetings, a portion of the conversation should cover the topic of maintenance. For example, if diesel generator loading will be 30% or more using building load, the building usually does not require regular load banking. The 30% loading is to ensure the generator's internal temperature will be hot enough to burn all of the injected fuel. Reaching high enough engine temperature avoids the build-up of soot in the exhaust stack. Soot is the product of light loading of a diesel generator for extended periods of time causing "wet-stacking." For permanent generators, one remedy for a wet-stacking concern may be built into the design by specifying a radiator mounted load bank, which is turned on by the generator on-board controller during periods of light electrical loading.

NFPA 110 notes unity power factor (purely resistive load banks) testing is permitted if the generator Manufacturer performed the nameplate rating test before shipment to the project site. If confirmations do not require reactive-resistive load banks, purely resistive load banks may be less expensive to rent for the on-site testing scope. For parallel generator installations, NFPA 110 only requires the anticipated running load to be simulated, not the entire parallel installation's combined capacity (kW).

Common characteristics to consider when specifying load banks:

- Many load banks have a multi-voltage rating. Using a higher or lower voltage may impact the load bank capacity (kW).
- Multiple compatible load banks may be controlled simultaneously by wiring their controllers together. For example, a 400kW block load may be possible by controlling four 100kW load banks.

- Load step quantity and resolution
    - For example, physical or digital user switches add load 0.7kW, 1kW, 6.5kW, etc. If the load steps are not precise enough, the testing strategy may be limited.
    - Suppose the Owner requires a load percentage for testing (exactly 100% of system capacity). In that case, the Commissioning Agent may need to scrutinize available load steps and capacity (kW) since the Owner may have a tenant contract to meet a particular percentage loading during commissioning. For example, many Owners find 80% to 95% acceptable for total loading, especially if the equipment was factory tested at 100% load. Otherwise, special attention to the available load steps and system pinch points (rating limitations) is needed to achieve exactly 100%.
- Determine if the equipment under test should supply the control power (fans and protection/control components) or if this control power needs to be from an independent source to avoid interruption during extended burn-in testing. If control power is from the equipment under test, note the fan and control power adds to the total loading.

<u>Temporary Cabling</u>

Cam-lock style connections bring value to projects without a high cost. It is recommended to consider their use for most designs. 2017 NEC Article 700 (emergency systems) requires a permanent switching means to connect to portable or temporary power. This switching means is also used when primary emergency equipment undergoes routine maintenance.

The cam-lock style connections utilize load bank cabling (Diesel Locomotive, DLO). These conductors are highly flexible, beyond a Class B or C conductor stranding, as defined in the NEC. If load bank pigtail cables are terminated in mechanical lugs, to maintain the lug listing, the conductor strands may need a copper or aluminum foil wrap (based on the conductor type) before terminating. Any double-lugs terminations should be rated if they are to accept more than one conductor. If connecting temporary load banks to I-Line style panelboards, they commonly require MLO plug-in molded case units for connection, as there are no exposed bus bars to connect to easily.

Standard load bank cable is typically 2kV rated (non-shielded) and available in #1/0 AWG and #4/0 AWG, using parallel sets as required for increased ampacity. This load bank cable, used in low-voltage applications, is commonly available in 50-foot lengths. Rental EPR cables are also available for medium-voltage up to 15kV, using either #2/0 AWG or 500 kCMIL. Medium-voltage cables are usually pre-tested with 200A load-break or 600A dead-break elbows. Medium-voltage load bank cables are also offered in 50-foot and sometimes 100-foot lengths.

When the Contractor runs multiple parallel sets of temporary cables, they should be arranged side-by-side (without separation) in the order: A-Phase, B-Phase, C-Phase, repeating for the parallel sets – avoid arranging cables A-Phase, A-Phase, B-Phase, B-Phase, C-Phase, C-Phase. Optionally, the Contractor may tri-plex these cables. Such cable arrangements seek to take advantage of canceling magnetic fields between the three-phases. Use caution for cable tripping hazards and consider keeping load bank connectors elevated above the ground if there is a concern for water in the area (rainwater or other Contractors may be filling mechanical systems with refrigerant or water).

It is prudent to monitor temperatures at connections using a point source heat gun. Avoid operating temporary load bank connections over their rated temperature (commonly rated 90C). Permanent terminations for low-voltage equipment are commonly 60C (100A and less) or 75C, and 90C for medium-voltage.

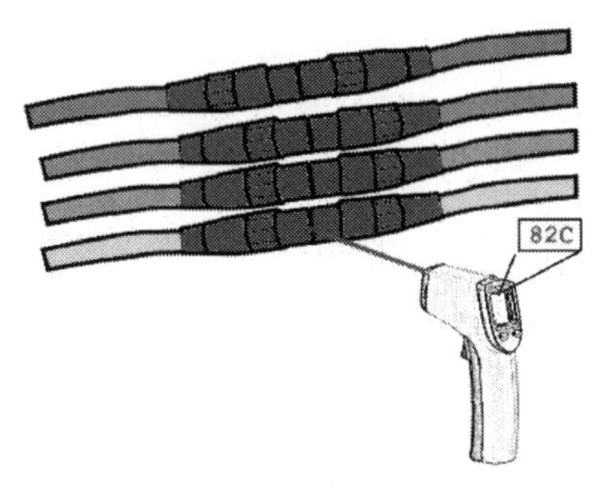

Figure 30: Point Measurement IR Gun Checking Load Bank Temporary Connections

Rack Mount Load Bank Cording

As part of the Test Equipment Plan, it is helpful to include the test load steps, such as 25%, 50%, 75%, or 100% of a total kW IT rack capacity. For rack-mount load banks, establishing the exact load steps may be tricky due to available per channel cording and rack power strip plug configurations. For the sake of load balancing, single-phase rack power strips (compared to three-phase strips) are especially challenging to coordinate the rack mount load bank cording. In some cases, rack mount load banks are physically mounted in one rack, but corded to the adjacent rack's power strip due to rating or cord limitations. Rack power strips are sometimes challenging to find detailed in the design documentation and are occasionally provided by the Owner rather than the Contractor. For proper coordination, the Commissioning Agent should request documentation for the specific rack power strip that will be installed.

To demonstrate the coordination effort, Figure 31/32 is just one of many possible cording configurations:

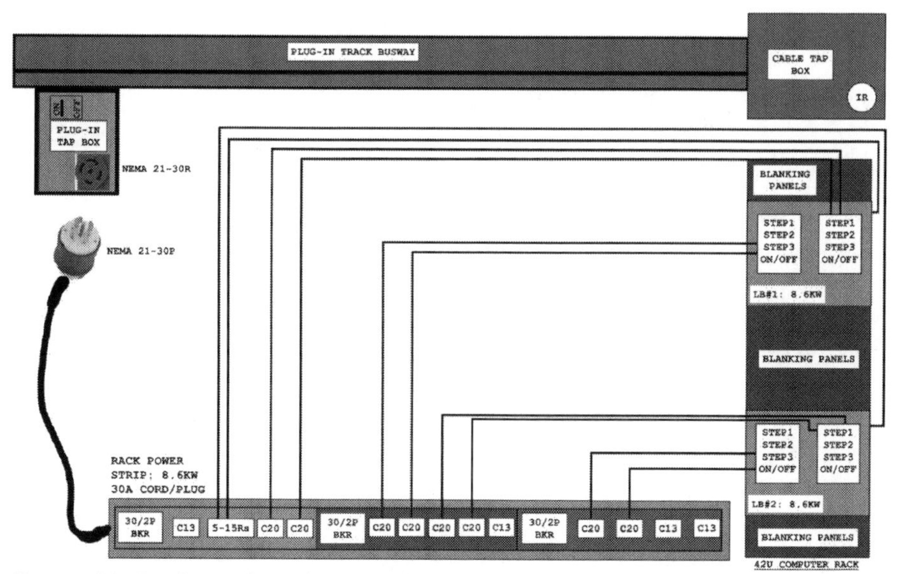

Figure 31: Rack Load Bank Example Connections

LB#	CHANNEL	STEPS	LINE 1	LINE 2	LINE 3
1	A	CNTRL	800	-	-
		ON	400	400	-
		OFF	[illegible]	[illegible]	-
		OFF	[illegible]	[illegible]	-
1	B	-	-	-	-
		ON	-	400	400
		OFF	-	[illegible]	[illegible]
		OFF	-	[illegible]	[illegible]
2	A	CNTRL	800	-	-
		ON	-	400	400
		ON	-	800	800
		OFF	-	[illegible]	[illegible]
2	B	-	-	-	-
		OFF	[illegible]		[illegible]
		ON	800	-	800
		OFF	[illegible]	-	[illegible]
			2800 WATTS (23A)	2000 WATTS (17A)	2400 WATTS (20A)

TOTAL 7200 WATTS (SHORT OF THE TOTAL 8600W CAPACITY)
20 AMPS (SHORT OF THE TOTAL 24A CAPACITY)

IEC 13 IEC 14
IEC 20 IEC 19
NEMA 5-15R NEMA 5-15P
CORD/PLUG TYPES

Figure 32: Rack Load Bank Example Loading

For the Figure 31/32 example, using C14 instead of C19 cords may make cording easier. C13 receptacles have a maximum rating of 12A and C19 maximum of 16A. The following spreadsheet may be helpful for rack mount load bank/power strip coordination (from Raritan, by Legrand):

https://www.raritan.com/blog/detail/how-to-calculate-current-on-a-3-phase-208v-rack-pdu-power-strip

Reference NFPA 70B, Annex I: NEMA Configurations. This is also a handy reference chart for receptacle configurations (PDF download):

http://www.inda-gro.com/IG/sites/default/files/pdf/nema.pdf

To complicate rack coordination further, as shown in the example (Figure 31/32), some rack mount load banks require a second cord (per channel) to achieve full load – this is due to the cable listing and as noted in NEC Article 400. The Electrical Commissioning Agent must carefully coordinate the rack power strips with the planned rack load banks. As the rack power strips and branch breakers are thermal-magnetic devices, they may trip at their rating or continue powering load for an extended time. As extended burn-in test failures may prompt re-testing, ideally, the load bank cording is well "cordinated" to avoid dropping load banks due to nuisance trips or overheating. When rack mount load banks overheat, the equipment over-temp light turns on and disengages the load bank. After cooling down, the load banks are available for use again.

Power Quality Meters

Power Quality Meters (PQMs) are also known as Power Quality Analyzers. Although some providers proceed with commissioning without PQMs, it is highly suggested to use them for electrical system confirmations. Load banks simulate building load, but the normal commissioning process is typically not using expensive electronics-based load banks. The basic load banks used for commissioning only go so far in mimicking the real load profile of

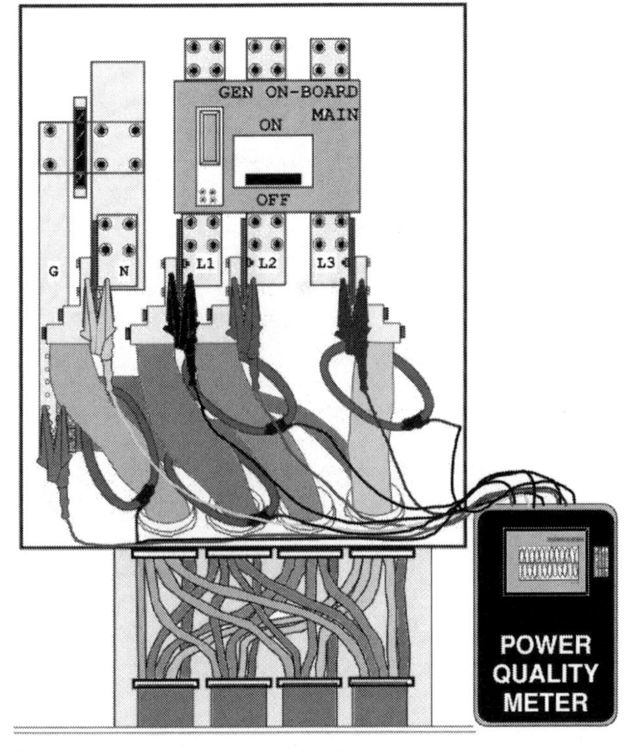

Figure 33: Generator PQM Connections

science instruments or computer power supplies that are commonly used in advanced building types. For this reason, PQMs bring value to the commissioning process. PQMs are particularly useful if the project includes critical equipment, such as UPS or generator systems. The portable meter has loose modular test leads and a basic digital display for local viewing; some screens are also graphic for real-time waveform observations.

One of your first questions may be: "Since my design already specifies permanent equipment mounted PQMs, may I use them for commissioning?" Maybe. Permanent meters do not always have transient settings (trigger limits). For example, to capture waveforms for block load steps, a portable PQM allows the setting of the transient trigger to capture an instantaneous snapshot of the load when it changes by, say 100A, or at specific load step percentages. Finding transient trigger settings in permanent meters can be challenging, which assumes that the particular meter has this functionality already. Using portable PQM meters for the commissioning process assures the user will have the desired features for testing. A portable (recently calibrated) meter is also helpful to compare against the readings shown by the permanent meters. Any permanent PQM used for commissioning should have uninterruptible control power.

The following are recommended characteristics to review when specifying a PQM for electrical commissioning:

- Quantity of:
    - Transient captures
    - Waveform captures
    - User waveform snapshots
- Historical trending (data logging)
- Wireless and/or ethernet/serial connection options
- Ability to set threshold alarms with start/stop times
- Type of data points: voltage, current, harmonics, crest, etc.
- Appropriate Category (CAT) ratings and product listing based on the voltage class
- Total memory: this is especially important if leaving the meter at the job site and for extended burn-in testing. Visit Manufacturer websites, as sometimes they make device memory usage calculators available by trend duration or anticipated quantity of transients/screen captures/alarms.
- Internal battery runtime: in case the control power is accidentally unplugged or lost during open transition system transfers

Product Examples

See the Rent Versus Buy: Power Quality Meters section. There are many PQMs on the market. When making selections, opening the user manuals is useful to understand the setup and functions. Examples:

- **Rx Monitoring Services:** http://rxms.com/index.php/products/meters/cx-monitor
- **Fluke:** https://www.fluke.com/en-us/product/electrical-testing/power-quality/438-ii
- **AEMC:** https://www.aemc.com/products/power-analyzers/power-8436
- **Megger:** https://us.megger.com/handheld-power-quality-analyzer-mpq1000-1
- **Dranetz:** https://www.dranetz.com/product/dranetz-hdpq-xplorer-400/
- **Hioki:** https://www.atecorp.com/products/hioki/pq3100

Manufacturer websites may provide web-based PQM comparison tools and some post PDF comparison literature. Optionally, try calling the Manufacturer's customer service and asking for a buying guide.

Leads and Connections

In addition to the PQM, the Test Equipment Plan should provide guidance on the test leads: IEC compliant CTs and voltage clamps. For example, the expected load current may be low enough to use simple clamp CTs, or a larger rope style may be required for higher amperage ratings. CTs are typically specified by noting the amperage. CTs come in many ampacity ratings such as 10A, 200A, 400A, 1200A, 3000A, 6,000A, and 10,000A. Never exceed voltage or current ratings for any connections – always de-energize, then read and understand the user manuals before a qualified personnel connects devices.

Since the Commissioning Agent will need to interact with meters during testing, longer test leads that allow the PQM to be located away from energized connections are preferred. Use cable sleeves to protect leads at equipment sharp edges. Identify if tests require CTs and voltage clamps on the neutral and/or ground connections. PQM meters may only have four CT positions, using the fourth input for either neutral or ground. The voltage clamps typically have both a neutral and ground inputs.

Take time to examine the PQM lead connections before energizing. If there are loose connections, the test will only take longer due to the repeat shutdowns and safety checks. Make sure CTs snap together and orient in the correct direction based on the right-hand-rule. Some CTs have arrows on them, which should be installed pointing at the load being monitored. Wrap the CTs around all parallel sets for each phase – missing a conductor in a parallel set will prompt troubleshooting. Once the meter displays steady-state waveforms, if magnitudes are excessive or in multiples, review the CT ratios settings. If total power does not add-up to the connected load, one or more of the CTs might be installed backward. Check voltage clamps or pigtail connections to ensure they are secure before energizing the equipment under test. Be sure the A-Phase voltage clamp is on the correct A-Phase feeder, and A-Phase CT is on that same phase - likewise for the other phases, or risk very confusing PQM data in the results. A more sophisticated meter may notice connection verification issues and prompt the user at the meter screen.

Many PQMs include an option to tether, via ethernet/serial or USB cable, directly to a computer for easy test setup. The tethered method is faster than using the PQM's local buttons for meter configuration.

Considerations

On the day of testing, after the Contractor installs the test equipment and powers up the system, the PQM meter configuration must be confirmed before functional testing may begin. The voltage should be visible on the meter, but without load (electrical current flowing), the CT connections will not be able to be confirmed. Depending on the test sequence, for instance a generator may require a cold-start, a small load step may help verify the CTs.

See NFPA 70B-Chapter 10 recommended PQM trending and trigger settings for capturing transient waveforms – this reference also has suggested points to be monitored. NFPA recommended settings should only be used as a starting place. Before functional testing begins, after a short sampling of data, the PQM configuration may require widening the sensing window (settings) for transients or risk running out of local memory during the test. Having excessively long user-prompted screen captures may quickly use up PQM onboard file storage also. Large data files will cause downloads of the data to take longer and may delay site testing. Ideally, PQM data is downloaded and confirmed after each test (while on-site).

Additional PQM considerations:

- **Use electrical tape and permanent marker:** Identify the PQMs for input/output, or PQM-1 and PQM-2, or rectifier input and static bypass input, etc.
    - The meters are used for multiple tests, so clearly identifying them is crucial.
- **Consider the computer peripheral connection type:** Serial or USB and any firmware drivers for converter cabling needed to interface with a laptop.
    - Optionally, some meters include wireless methods.

- **Make sure to set the time clocks for all meters:** Times between meters should be the same. Having clocks synchronized aids in correlating and troubleshooting data.
    - Some PQMs allow the user to "push" (synchronize) their local laptop time to the meter so that both meters have the same clock.
- **Review PQM Software:**
    - All PQMs have Manufacturer-specific (free) software
    - Some software packages are friendlier than others
    - It is beneficial to download and overview the software. Some provide a sample data set for experimental use. The Manufacturer may even have a meter simulator to download and experiment. For example, the software may generate CBEMA/ITI curves. Other curves to consider: IEC 6240-3 (curves 1 through 3, representing sensitive to general-purpose IT loads).
    - Almost all Manufacturer PQM software options have separate applications for configuring the meter and viewing saved data
- **Some PQMs include smartphone apps:** In combination with longer test lead lengths, this is a safety measure for the user to keep a distance away from energized equipment
- **Be diligent in noting the time and activities:** For example, 7:46AM - 50% step load on UPS-2. Having this detailed written record is the key to navigating the PQM data at the end of testing. Some meter software also allows digital note-taking for steps, but a handwritten log is strongly recommended for back-up.

Infrared Camera

Infrared (IR) scanning is ideally completed by a Certified Thermographer. If specified by the design, IR ports (integral equipment windows – available in various sizes) are a safety forward approach to scanning. The equipment will need to be energized and under load (demanding current). If incident energy is over 40 cal/cm^2 and the setup must have equipment doors opened during scanning, before proceeding, refer to NFPA 70E.

Standard Preventative Maintenance (PM) programs seek to monitor terminations and investigate relative temperature differences between phases. Reference NETA and NFPA 70B-Chapter 11. Comparisons are between similar components that are under similar loading, as well as compared to the ambient air temperature. Depending on the difference in temperature, recommended actions range from investigation, repair over time, continue monitoring, and immediate repair.

The Owner may prefer a special format for recording data, in case they intend to trend the baseline data along with their interval maintenance over time. Keep in-mind, any electrical current un-balance will be seen as more heat on the phase with the highest demand current, so it helps to record percentage loading or actual current on each phase for reporting. Any significant discrepancies are reviewed by de-energizing, performing live-dead-live/LOTO, and checking for proper torque at terminations.

The final commissioning report should include an appendix for the IR report. Even if the scans do not find concerns, a handful of regular photos side-by-side with infrared scans are useful for the record and the report narrative; this is ideal instead of a report with only narrative and no baseline temperature data references. Without the normal photo, the infrared photo may not be distinguishable.

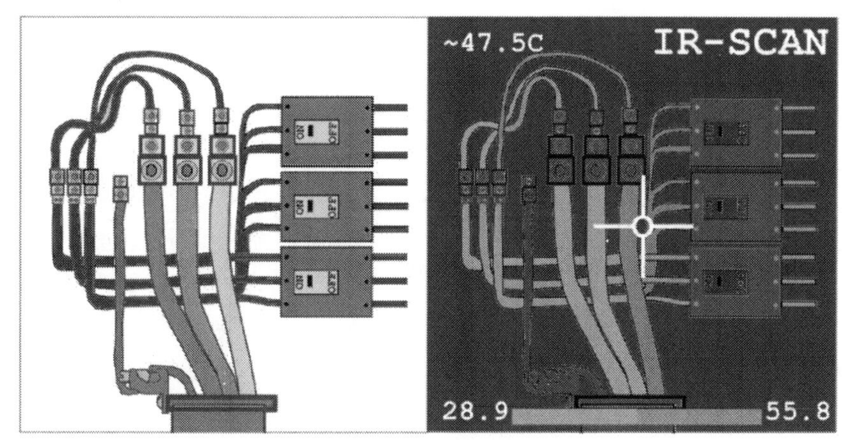

Figure 34: Side-by-Side Normal Photo and IR Scan

Additional considerations for IR scanning:

- How long does a circuit need to operate at full-load before it has reached a steady-state heat? Short of an industry reference, one-hour is a reasonable metric for the sake of IR scanning. If one monitors the terminations over time a steady-state temperature should become apparent also.
- The scanning process should review the entire piece of equipment and circuit terminations, particularly the field-made terminations. The components scanned are not only mechanical terminations. For example, it may make sense to IR scan large circuit boards containing electronic components. It helps to pre-plan scanning before starting testing, so the person scanning and the Commissioning Agent agree on the objectives.
- Medium-voltage equipment, such as metal-clad switchgear, usually requires insulated boots at mechanical terminations to meet equipment listings. It is recommended to leave these boots installed during scanning. Although the boots do impact the temperature readings, they are relative differences and consistent with scans that will occur during maintenance.

Other Test Instruments

Additional portable instruments that may be listed in the Test Equipment Plan are suggested below. A hands-off commissioning role should work with the Installing Contractor to perform any activity beyond de-energized equipment inspections.

- Flashlight
- Rotation meter
- Continuity tester
- Live-voltage indicator
- Partial discharge scanner
- Basic receptacle tester
    - Include GFCI test function
- Low-resistance ohm meter
    - Used to confirm bolted connections
- Electrical multi-meter
    - Practice live-dead-live meter techniques
- Internal ohmic battery tester/analyzer
    - Used for some battery systems
- Light meter
    - Used to verify the emergency and normal light levels
- Sound meter (or phone app)
    - To confirm equipment sound levels against specifications
- Humidity, temperature, or similar rental sensors
    - Coordinate with the Mechanical Commissioning Agent to confirm
- Simple point IR measurement gun
    - Review equipment and temporary connections to confirm they do not operate beyond their product listings

When electrical acceptance testing is already specified in the design specifications and not required by the Owner, the Test Equipment Plan excludes acceptance testing equipment. For example, a DC High Potential (DC-Hipot) is one option the Design Engineer may specify for acceptance testing new extruded medium-voltage cables.

All measurement instruments require calibration, typically every twelve-months. The firmware should always be up-to-date (if applicable). If the device calibration sticker (decal) and the certificate has exceeded twelve-months, refer to the National Institute of Standards and Technology (NIST) to evaluate the appropriate calibration interval in more detail. The calibration certificate might be printed or digital on a USB drive stored in the PQM shipping container. Be sure to copy any digital files to a company laptop before returning the meter. These certificates should be included in the final commissioning report.

Planning

As noted in the Project Management chapter, there is a significant cost associated with test equipment. The Test Equipment Plan must be clear on who is responsible for test equipment rentals (Contractor or the Commissioning Team). The Commissioning Team should prioritize the Test Equipment Plan document and make it available for the Contractor Team members to review and assign costs as soon as possible. Knowing the exact cost for commissioning, including equipment rental costs, helps the Owner understand the remaining project budget for other priorities.

The Test Equipment Plan is also used to document test schemes. For example, planned testing may require re-cording of load banks to different rack power strips to achieve testing goals. This plan shows where the load banks and measurement instruments interact with the electrical system. One suggestion is to mark-up an electrical one-line diagram and identify the connections to be on the line or load side. A floor or site plan mark-up may also be needed. Coordinate the connection methods with the Electrical Contractor to leverage their expertise and determine if they have another connection location preference. Generalizing how the equipment will connect to the system leaves unknowns that may cause project delays. For example, to accommodate the required load, the Contractor may need multiple load banks with a significant amount of temporary cabling to connect to the permanent distribution system. By not having a clear plan of how to connect to the electrical systems, safety planning for the activity may also be hindered.

The Test Equipment Plan provides insights to aid overall project coordination. This plan prompts all involved parties to consider, "How are we going to implement a successful testing strategy?" The outcomes of these determinations then impact the construction budget and schedule. With the level of importance that the Test Equipment Plan brings, many stakeholders are interested in reviewing it. As such, this is an excellent document for the Commissioning Agent to write in coordination topics.

For instance, to maintain the desired heat distribution for IT racks, the Commissioning Agent may have particular rack-U locations where the rack mount load banks should be installed. The plan may show an elevation of the racks with load banks positioned and blanking panels to fill the gaps. Blanking panels are necessary to test data center hot or cold aisle air containment systems. Before renting blanking panels, check with the Owner to see if they already have blanking panels that the Construction Team may borrow for testing. Fabricating temporary blanking panels on-site is not recommended. By writing about the blanking panels in the Test Equipment Plan, the team is made aware of the need for coordination.

Depending on the testing configuration, additional temporary components may be needed. For instance, if attaching load banks to busway, then a rental tap box may be required and the permanent busway make/model may be needed. The Contractor's Vendor may fill-in these details, but it is helpful to highlight accessories directly in the Test Equipment Plan if possible.

In existing installations, the Contractor or Commissioning Agent may need to submit an IT outage request to the Owner's Technology Team. The request may require several Management Staff signatures to approve the outage, so the Test Equipment Plan is an excellent place to outline the Owner's required procedure and timeline.

Here are a few other coordination examples for the author of the Test Equipment Plan to consider:

- Confirm if the Owner's infrastructure will be available for testing: IT racks, power strips, tap boxes, blanking panels, etc.
- Load bank testing of scale is a loud activity. Verify if the building has local ordinances for noise. Also, consider if Contractors may be in the close vicinity trying to perform other work on electrical functional testing days.
- Preplan the outdoor load bank location to avoid heating obstructions and causing damage or fire.
- Investigate separate control power connections and build-in safety features for equipment ahead of time. As an example, during significant generator load steps, the voltage may dip more than 30%. If the load bank's fan control power is integral (not independently powered), the generator fan power may drop-out. The lack of fan power causes the load bank's output contactors to open in order to avoid overheating the internal elements.
- Consider providing temporary labeling for power, control wiring, and other electrical connections. Label the branch breaker if the load bank has separate control power – this may save from other electrical work occurring on that circuit occurring during functional testing. Such an event could interrupt an extended burn-in test and cause a re-test requirement.
- Highlight the need to remove load bank covers and confirm the fan spins in the correct direction before testing starts.
- Note additional test equipment for acceptance testing outside of the commissioning scope may be needed. For test procedures that may inform additional acceptance/maintenance testing equipment, reference NECA 90, Annex A, and NETA.
- Identify the potential scope for modifying equipment settings. For instance, the commissioning period is an appropriate time to consider the right transformer tap settings since the building voltage experiences both no-load and full-load.
- Provide appropriate grounding and bonding for any trailers holding test equipment.

- Consider charging batteries for at least seventy-two hours before functionally testing them during the commissioning process. Using this approach makes sure the batteries have a full charge for a real timed test during discharge. As noted in the Pull Planning section, air conditioning should also be available for battery functional testing.
- Comment on field conditions that may impact test results. For example, a thermal scan through a plastic shield may alter the results. Comments such as this in the test plan prompts the Electrical Contractor to consider how they plan to complete this scope.

The final Test Equipment Plan must be reviewed by the mechanical and controls Commissioning Team members, as the specific testing schemes are often driven by what the Mechanical Team needs to prove with their systems. Call your local test supply Vendors, who are happy to assist and likely provide budget numbers for estimating use.

Chapter 7: Test Scripts and Checklists

General Guidance

In the Commissioning Process chapter, we detailed Levels 1 and 2 of the construction phase. We also briefly touched on Levels 3 through 5 and in this chapter, we provide further detail. Level 3 IVCs, Level 4 FPTs, and the Level 5 IST are all documents written by the Commissioning Agent. The Installing Contractor completes the IVCs. As a consultant in a witness only testing role, the Commissioning Agent usually completes the FPT and IST scripts.

Typically commissioning software platforms create digital timestamps for all entries. For example, if a test script step has a *Yes/No* input, when the user presses *Yes* or *No*, the program makes a timestamp (date/time) including the user who created the entry. Suppose the Commissioning Team writes scripts manually, via spreadsheets or word processing programs. In that case, the best practice is to include a column adjacent to every test step for the tester's initials/date/time. Optionally, instead of *Yes/No*, the test steps may prompt the user to record values or take a picture of a test instrument or the equipment under test. As best as possible, every test step should have definitive pass/fail criterion without having to look at other references. The final commissioning report includes both the successful and failed test attempts. Each checklist or test script should also consider noting the outside weather or room conditions (temperature/humidity).

<u>Level 3 – Installation Verification Checklists (IVCs)</u>

Level 3 starts at the equipment delivery stage. If included in the project scope, the Contractor begins completing IVCs for equipment upon delivery to the project site. For example, after unboxing and performing general checks for loose parts and damage, the equipment nameplate data is documented. These checklist items may be included in the first section of the IVC. The Contractor should compare the equipment nameplate data to the design documents and the EOR approved Contractor submittal. For the project record, overall photographs of the arriving equipment are also recommended. Using the equipment delivery as a step in Level 3 helps the Manufacturer address issues quickly. For example, as soon as a transfer switch arrives at the project site, it is best to confirm if three versus four-pole (switched versus un-switched neutral) was delivered.

The equipment manuals are a significant source of information to understand proper site verifications. Ideally, the Commissioning Agent reviews the equipment installation manuals and considers those requirements when making Level 3 checklists. The IVCs are not a substitute for the Contractor reading the installation manuals and following the Manufacturer's written instructions, but a means to double-check the installation. Quality electrical IVCs also lean on industry standards, such as NECA, NEMA, NETA, IEEE, and NFPA 70B. For example, NFPA 70B includes dedicated chapters for the most common types of electrical equipment. The product listings are yet another source of information to build install checklists. See Nationally Recognized Testing Laboratories (NRTLs) to search equipment listings:

https://www.osha.gov/dts/otpca/nrtl/nrtllist.html

If the Contractor has comparable installation checklists that they prefer to use for this stage, the Commissioning Team may evaluate and determine if they are appropriate. For example, the Contractor may have a standard template for torque logs. The IVCs also serve as a location to record information such as the equipment-controller default passwords. NFPA 70B, Annex H forms are a great source of information as well. Another reference for inspection checklists (PDF download):

https://electrical-engineering-portal.com/res3/Electrical-Inspection-Checklists.pdf

IVCs aid in the Contractor-to-Owner construction handover of the construction documentation. Building Owners sometimes use software for maintenance data collection. For example, the software package Construction Operations Building Information Exchange (COBie). If the Owner has specific software for maintenance, the Commissioning Agent works with the Owner to understand the fields required for each piece of equipment, such as model, manufacturer, location, voltage, capacity, etc. The Commissioning Agent includes these fields directly in the IVCs for the Contractor to log data from the facility equipment as they are verifying the installation. Using a cloud-based software solution allows the IVC data to be extracted quickly and shared with the Owner Team for their maintenance software. The Owner Maintenance Staff might walk the building after construction and add barcodes to the equipment; this integrates the equipment into their maintenance program for quick access. Optionally, the data may be available via the Design Team's Building Information Model (BIM). These maintenance platforms may also provide data for import into a Computerized Asset and Facility Management (CAFM) or Computerized Maintenance Management Systems (CMMS).

Once the Contractor has completed the IVCs' installation portion, the equipment start-up portion of the IVCs may occur. As required by the contract documents, after acceptance testing, either the Contractor or the Manufacturer Representative performs start-up. If the Contractor completes the start-up, they should follow the equipment installation manual and Design Engineer Part 3 specified instructions.

Mike Starr, PE

Some of the most frequent checks performed before energizing and configuring the equipment include:

- Electrical phase rotation is correct
- Ventilation fans move unobstructed
- Branch heater wiring connects to the proper voltage
- Mechanical and electrical interlocks are functioning as required
- No debris exists after the Contractor has completed the installation

See the Equipment Inspections section for further insights.

The Commissioning Team commonly requests that the Contractors complete all Level 3 IVCs before Level 4 testing begins. For example, Level 3 checklists will have the Contractor confirm network connectivity by visually confirming blinking lights on Network Interface Cards (NICs). Having incomplete Level 3 checklists may generate issues when the Commissioning Agent arrives on-site to start Level 4 functional testing.

Level 4 – Functional Performance Testing (FPT)

Level 4 involves Functional Performance Testing (FPT). On-site activity for the Commissioning Agent increases at this level, as they now witness test any equipment that has a functional aspect. Test scripts typically start with high-level installation checks, somewhat in a spot-check fashion compared to the Contractor completed Level 3 IVCs. The Commissioning Agent attempts to limit Level 4 install checks to critical details. If spot-checking finds issues, more detailed inspections are needed. Confirmation of the equipment control and monitoring points occurs during Level 4, as they may not have been available during Level 3. Checking points involves interfacing with the Building Management System (BMS); thus, Mechanical Team coordination is usually needed to synchronize these events. The data center industry may also use a Data Center Infrastructure Management (DCiM) system, which is another source to verify electrical points; likewise, some applications include Process Control Systems (PCSs).

The root of Level 4 involves performance testing equipment and combined systems. In some cases, a Sequence of Operation (SOO) fully describes the system, but drawing a state-diagram may also be useful to ensure all modes of operation are tested. For example, normal, abnormal, maintenance, emergency, economy, sleep, restart sequence, emergency power off, etc. The Design Engineer's SOO explains the intent, but the sequence to utilize for the commissioning effort should be based on the Manufacturer Application Engineer's official shop drawing sequence – the EOR should have confirmed these sequences during the CA phase. All failure modes need to be simulated – not just the documented failure modes, but also undocumented instances the

Electrical Commissioning Agent recommends based on their experience. All functional testing should be repeatable to confirm the equipment's function is predictable.

Separate from the commissioning Levels 1-5 convention, NFPA 110 and 111 describe Level 1 and Level 2 systems. Note: for the NFPA standards, Level 1 systems have the potential for loss of human life or serious injuries. Generators and lighting inverters are often Level 1 systems, and UPSs are commonly Level 2 systems. For inverter systems over 500VA, Level 1 systems may require discharging the DC link for fifteen-minutes or up to the rated battery runtime. Per the code requirement, the AHJ must be given notice to witness the on-site testing for Level 1 systems. Local codes may have further requirements beyond NFPA 70 (NEC): 700 Emergency Systems, 701 Legally Required Standby Systems, 702 Optional Standby Systems, and 517 Healthcare - Life Safety Branch, Critical Branch, Equipment Branch.

Below is a summary of NFPA 110 generator testing, but the Commissioning Agent should perform due diligence by comparing this against the standard directly:

- Cold start
- 1.5 hours paralleled with the expected load
- 5 minutes of idle
- Each generator:
  - 30% load for 30 minutes
  - 50% for 30 minutes
  - 100% for 60 minutes

For parallel UPS and generator systems, NFPA may require parallel testing to occur first. If the team prefers to proceed with individual generator testing before the parallel test, it is best to confirm this with the AHJ ahead of time. Parallel systems then have the parallel FPT and also an FPT script, per unit.

If the Owner Maintenance Staff members are qualified (knowledgeable and experienced) to operate the electrical installations, some Commissioning Providers conduct pre-test training before Level 4 testing begins. Doing this helps massage finer test scripts details and prepares the Owner Maintenance Staff to operate their new equipment.

See the Load Requirements section for further insights.

Mike Starr, PE

Cold Start

For generator systems, a cold start is when the engine has not been running but is also not completely "cold." Instead, the generator is as it would usually sit on-site. That likely means unit heaters and block/alternator/battery heaters should operate on cold days. When transfer equipment signals for generation to start and carry the building load, the cold start test attempts to simulate the actual conditions to see if the back-up source can support the connected loads during a normal power outage.

Traditionally, the building loads are powered by generation in priority steps, controlled by conditions such as transfer inhibits, programmed priorities causing transfer equipment to connect/disconnect, and/or control systems opening/closing remote breakers. Designers utilize free generator consumer software to confirm the generator sizing based on the site conditions and load characteristics.

Step and Block Loads

Although NFPA 110 and 111 do not explicitly require step and block loading in the manner described below, this testing is short duration and helpful to perform prior to starting an extended burn-in. Step and block loading may root out significant issues before spending hours stressing the equipment at constant load.

Step loads are incremental as a means to exercise the equipment before attempting block loading. For example, for a 50% step load, the load bank operator leaves the master switch in the on position and then increases the load until the 50% target (such as 250kW of 500kW capacity). An FPT may require loading in steps: 0%, 25%, 50%, 75%, and 100%. Steps may add or decrease load to test power electronics and unloading of generation. For example, steps: 100%, 75%, 50%, 25%, and 0%. Approaching the steps loads in this manner, and allowing one to two-minutes pause at each step, makes the data easier to locate a mountain shape in the PQM data. The Commissioning Agent may also assign step loads that closely align with modular equipment building blocks. For example, some UPSs have power modules in 25kVA up to 75kVA range, which could be a metric to consider load percentages that cause power modules to share load or cycle for economy.

Block loads are meant to be nearly instantaneous, demanding the equipment under test support load quickly and without interruption. For example, for a 75% block load, the load bank operator would configure the load steps to equal 75% of the total load, then turn the master switch to the on position. One of the reasons for block loads is to judge the system transient response as it reacts to a dynamic load change. When the block load occurs, there may be a significant voltage dip. Also, when the load is disconnected, the equipment control systems must maintain voltage and frequency stability. For example, an FPT may require: 0% to 50%, 50% to 90%, 90% to 0%, 0% to 100%, and 100% to 0% block loads.

An example of a 300kVA UPS having step and block loads is represented in Figure 35:

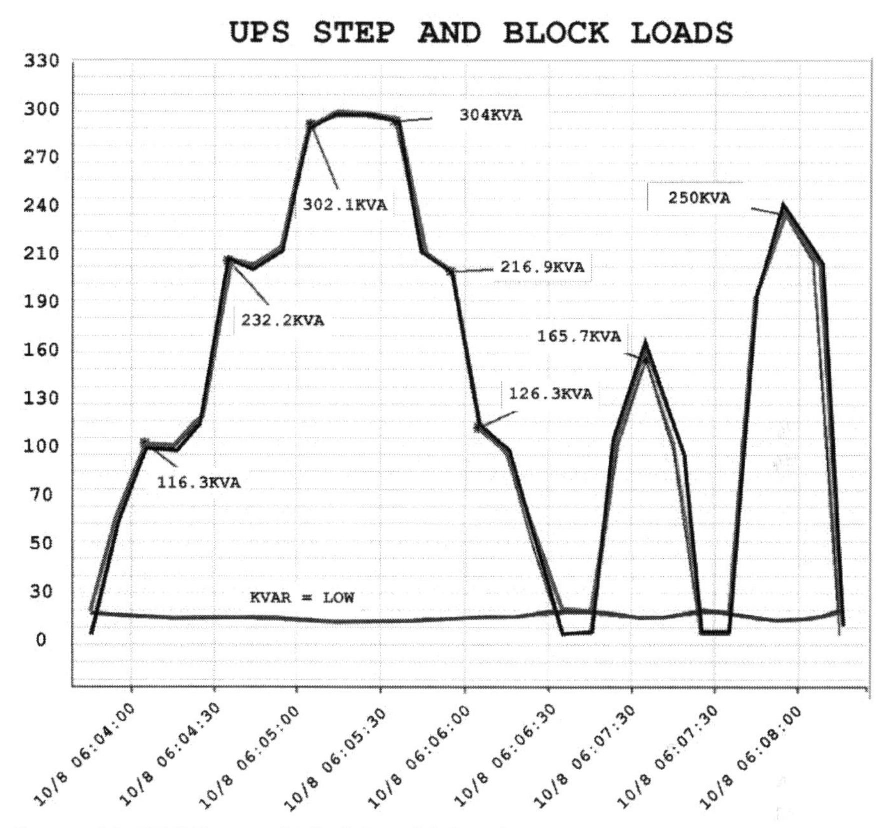

Figure 35: UPS Step and Block Load PQM Output

Extended Burn-In

Now for the number one question: "How long should the extended burn-in portion of the test be (length of time)?" Extended burn-in happens after the equipment goes through sequences for functional testing. Burn-in is the very last activity on the Level 4 script. For example, a generator system may continue to power load banks at periodic, constant load steps for multiple hours. For UPS and generator systems, see NFPA 110 and 111 for Level 1 code mandated load steps and burn-in times.

Outside of a code requirement, extended burn-in times range from one-hour up to twenty-four hours. Around two to six-hours is a typical burn-in time. The length of time for testing depends on the criticality of the application.

If the Owner makes the investment to have equipment and personnel with expertise on-site for processes that include extended burn-in testing, there is value in making sure the equipment is up to the job of supporting the building application. Taking this opportunity to truly understand the boundaries of the equipment performance is a valuable exercise. Once the team and equipment are on-site, hooked-up, safety checked, and running, a couple more hours of testing may not be significantly more expensive. Unmanned sites are especially good candidates for considering longer burn-in times.

Failure Modes

The Manufacturer shop drawing and equipment manuals should list failure modes. In addition, the Commissioning Agent should review the design specifications to see if the EOR has more lenient or extreme testing requirements. For example, checking the system response for these types of failures:

- Loss of DC link
- Door open alarm
- Simulated fan failure
- Loss of control power
- Loss of communications
- Back-up source fails to start
- Protective device fails to open/close/trip
- Extended parallel conditions prompting lock-outs

During Level 4 FPTs, the Commissioning Agent should work closely with the Manufacturer start-up technician to see if they have additional failure mode checks that may be performed.

Level 5 – Integrated Systems Test (IST)

The IST is a coordinated effort to see all building systems work together. All major systems now have the opportunity to demonstrate redundancies and maintain the building loads through a series of test schemes. The IST is also known as the pull-the-plug, lights-out, or black-start test. These names come from the most significant activity in an IST sequence, when a normal utility loss causes the building standby systems to initiate. Ideally, the actual normal power source is de-energized, but pulling sensing fuses may also be an acceptable way to simulate a power outage. Do not forget to put the system, and all master-slave controllers, in automatic mode first! The IST is also the time to perform building-wide EPO testing and, more importantly, confirm the exact sequence required to bring systems back to normal conditions thereafter.

Although the commissioning process includes Levels 1 through 5, skipping straight to Level 5 testing may be appropriate for a small non-certified project. Ideally, all building trades (electrical, fire protection, security, etc.) have successfully completed their acceptance testing (Level 2), install checklists (Level 3), and functional testing (Level 4). It takes a significant amount of pre-planning to have a successful IST. On the day-of, Contractors from every major building trade should be on-site and ready to assist – this significantly helps issue resolution and avoids schedule delays.

Before starting Level 5 testing, the commissioning Issues Log is complete or has very few in-progress items remaining. By implementing the commissioning process, the IST is hopefully completed with no further or only minor discrepancies. In some cases connections were not under load during previous testing steps and the IST is the final opportunity to Infrared (IR) scan equipment. For example, the mechanical system may not have been running when functional testing happened for corresponding transfer switches (no-load functional testing). If the calculated incident energy is less than 40 cal/cm^2 and the thermographer wears appropriate PPE, the transfer switch terminals may be IR scanned to review for loose connections – see NFPA 70E.

By the time the Project Team attempts the IST, any misalignments in transfer switch and building automation settings for the timing of transfers will become readily apparent. For example, in a data center, waiting ten seconds for a generator to start, transfer equipment to change sources, and the power to be restored to the cooling equipment is a long time for computer equipment (backed-up by battery power) systems to continue generating heat without cooling. If the entire building system is not designed and calibrated correctly, the supporting cooling equipment may not recover the room thermals. For IT cabinets, as much as the Owner would like to advance their process by installing rack equipment, for the sake of protecting the Owner's equipment, it is best to wait until the IST is complete.

Level 5 is the opportunity to test the lighting system. Recall lighting design considers depreciation factors, such as dirt, and also assumed room surface reflection percentages. Normal light levels may need to dim to a lower light level for the Day 1 settings. Light measurements are made at the application's working height. A classroom may have normal lighting measurements thirty inches above the floor at the student desk height. In an egress corridor, emergency lighting measurements are made at the floor – see the Life Safety section.

Specific Electrical Systems

For this guide to function as a reference, the following sections highlight considerations for commissioning specific electrical systems. Insights from early discussions in this chapter are the basis. Potential sub-systems that may have testing needs are also listed. In context with the full scope of this guide, the pointers for specific systems seek to generate Project Team conversations about testing for various building applications. Although scripts and checklists are not provided explicitly, there is hopefully enough context and suggested outside sources to develop IVC/FPT documents from scratch or to improve your company master scripts.

Given, the same electrical systems identified in the remainder of this chapter have additional types. For example, some generators have alternate fuel options (dual fuel, dynamic gas blending, etc.), and there are even more unique applications, such as Combined Heat and Power (CHP). There are also many configurations for parallel UPS static switch arrangements. Likewise, there are variations involving physical generator paralleling equipment versus paralleling with on-board control (contactor-based solution) and remotely opening/closing breakers. This guide's suggestions reflect the most common installations rather than covering all design possibilities.

As noted in the Introduction, the solution to a commissioning problem may originate in the design phase, so the considerations in this chapter also contain a balance of design discussion.

Application-Specific Considerations

Installations may be subject to additional requirements based on their applications, such as healthcare (NFPA 99, NEC 517, Joint Commissions, etc.). For example, the Owner may request a more detailed commissioning effort for an in-patient (I-2 Occupancy) than an ambulatory building (B Occupancy). In the mission-critical market sector, The Up-Time Institute certification compliance may require a Data Center Continuous (DCC) rating for the on-site generation. NEC Article 700, as of the 2017 NEC, requires maintenance and testing for the emergency power system. Government buildings have their own set of standards that must be complied with; likewise, each market sector has specific requirements.

The following is a non-inclusive list of codes and standards that range in application. They should be referenced in the process of designing and commissioning of building electrical systems based on the application:

International Building Code (IBC), Code of Federal Regulations (CFR), American Society for Testing and Materials (ASTM), National Electrical Manufacturers' Association (NEMA), Illumination Engineering Society of North America (IESNA), Americans with Disabilities Act (ADA), Occupational Safety and Health Administration (OSHA), American National Standards Institute (ANSI), Institute of Electrical and Electronics Engineers (IEEE), National Electrical Safety Code (NESC), Environmental Protection Agency (EPA), Federal Emergency Management Agency (FEMA), Office of Statewide Health Planning and Development (OSHPD), Department of Energy (DOE), Owner insurance, energy codes, equipment listings, sustainability standards, seismic, etc.

Confirming all applicable codes and standards are applied is the responsibility of the Design Team. The Installing Contractor must also comply with the Manufacturer instructions. Consider having qualified acceptance testing for all installations. To confirm if prescribed testing is required, the Commissioning Agent should review the necessary standards based on their specific building application.

Electrical Distribution Equipment

Some power distribution equipment does not have functional aspects. For example, panelboards have opening and closing circuit breaker verifications, confirming circuits are installed as they are labeled, metering, and surge protection type verifications. Much of the panelboard functional checks are for the monitoring points related to the BMS. Buildings with sustainability design features involve metering confirmations for the Commissioning Team, such as those used in total building energy summary screens. Unless they include a local controller or gauges for monitoring, transformers also have minimal Level 4 functional requirements.

To confirm their custom design solution, specialty equipment with automated schemes are common candidates for functional verifications. Testing for such equipment may include: load banking, cycling source equipment on-off depending on load demand, load shedding/pick-up, paralleling, etc.

Depending on the complexity of the equipment, consider confirming:

- Relay protection schemes
- Enhanced safety mechanisms, such as:
    - Fiber-based arc flash detection – tested by flashlight
    - Zone Selective Interlocking (ZSI) (arc mitigation strategy)
- Polarity for DC systems, such as station power
- Returning A source first, then B source, and try the opposite to observe if there are anomalies

- Rotation is critical for UPSs, generators, motors, and transfer equipment. See Phase Sequence, Rotation, and Arrangement for further insights.

See the Equipment Inspections and Field Observations sections for further insights.

Life Safety

Life safety equipment is often witnessed by the Authority Having Jurisdiction (AHJ). Thus, other than generators and lighting inverters, emergency aspects of lighting (egress light levels) and the fire alarm system are sometimes excluded from the commissioning scope – since the AHJ will be overseeing this scope as an occupancy requirement. Fire alarm system verifications are heavy on wiring verifications (points). Gaseous suppression fire protection systems require Contractor due diligence in sealing all room penetrations to meet the NFPA prescribed room pressurization.

The Authority Having Jurisdiction (AHJ) may require a walk-through of the building to confirm emergency lighting levels. Upon loss of power, the dimmed emergency lighting fixtures must go to full-bright. Confirm emergency lighting levels: 1.0 foot-candle (average), minimum of 0.1 foot-candle at any point, and a maximum-to-minimum maximum ratio of forty-to-one. The light levels may depreciate by 60% by the end of the 90-minute test. For egress (emergency) lighting, light measurements are always measured at the floor. Confirm the lighting control system head-end equipment has the appropriate power (emergency versus normal) as described by the product manufacturer.

Lighting Control

Lighting control can be complex. Multi-day courses are dedicated to the commissioning of lighting control systems. Designs range from line (V_{L-N}) or low-voltage addressable light fixtures controlled by centralized or local room controllers. Finding a simple relay panel design (without dimming control) in today's advanced buildings is becoming less common. Color changing LED is also a common practice now.

The lighting sequences/set points/schedules are the types of considerations for commissioning the lighting system. The Commissioning Agent will also need to review the user interface setups (local switching schemes). Light levels for spaces listed in the Design Team BOD are also confirmed. In addition to building codes, reference NFPA 101 and NFPA 70B-Chapter 23.

Wiring Devices

General-purpose receptacles are without significant functional testing beyond a basic receptacle tester and confirmation of their branch circuit source. A plug-in receptacle tester allows the testing of GFCI protected receptacles, as well as checks for open neutral/ground/hot and reversed connections. Other functions required by energy code, such as controlled receptacles, may need to be verified. Automated shade controllers fall into the device level functional testing also.

Generators

See the sections earlier in this chapter for further insights.

Brief Description

This section considers generation, packaged in an exterior enclosure, or in a dedicated interior room. The fuel storage could be a sub-base (belly) tank or a remote above/below ground main tank with local day tank. Gensets have various emission tiers and are available in standby, prime, and continuous ratings. Tier 4 Final is required for unlimited non-emergency use (loss of utility). This guide recommends reviewing generator supplier white papers to gain a better understanding of emission ratings. Optionally, the fuel source may be piped natural gas. Generators are 0.8 power factor rated. They might be paralleled for higher capacity. Paralleling is either via onboard generator controllers and remotely opening/closing breakers or with dedicated distribution equipment containing synchronization check relays. Generators are common for both low and medium-voltage applications. Generators serve NFPA 110 Level 1 emergency loads or Level 2 optional standby loads. The NEC designates these loads based on Chapter 7 and Article 517.

Abbreviations

Gen, Genset (Generator Set: Combined System)

Types

Diesel, Natural Gas

Codes/Standards

In addition to the notes in this chapter's earlier sections: NEC 445, NEC 480, NEC-Chapter 7, NECA 404, NFPA 30, NFPA 37, NFPA 110, UL, EPA, and ISO.

There may also be mechanical ventilation and fuel-related considerations. Coordinate closely with the Mechanical Commissioning Agent.

Key Functional Tests/Verifications

Step loads	Block loads
Monitored points	Emergency Power Off (EPO) and manufacturer safeties
Settings confirmations	Load pick-up/shed sequences
Cold start	Automatic load bank dump if called to start
All modes: automatic, manual, load banking, temporary roll-up, and failure modes	Extended burn-in
Repeat all modes: while in parallel, if applicable	Potentially an emissions test
NFPA 110 requirements for Level 1 systems. There are factory tests required to avoid site inductive load bank testing. Some tests may be difficult to simulate on the project site, such as the crank test.	

Table 4: Generator Functional Testing

Measurements

Electrical trending	Sound level
NFPA 110 Level 1 system annunciator required points	Equipment total power consumption from the local controller, PQM, and load bank
Battery and battery charger voltage	Fuel consumption
Time it takes to transfer for comparison against NEC-Chapter 7 requirements, if applicable	Infrared (IR) scanning is not normally completed on generators due to excessive incident energy
Oil temperature	Oil pressure
Coolant temperature	Exhaust temperature
Transients during step loads	Transients during block loads

Table 5: Generator Test Measurements

Sub-Systems

Remote EPO	Vibration isolators
Network card	Grounding resistor
Space heaters	Transfer equipment
Leak detection	Remote annunciator
Stair platforms	Reverse power relay
Battery charger	Lightning protection
Fill alarm panels	Fuel monitoring system
Polisher systems	Best battery selector system
Diesel particulate filters	Fuel pumping systems
Remote radiator systems	Generator branch power distribution
Outdoor drop-over or sound attenuated enclosure	Batteries, block, and alternator heaters
Fuel tanks	Radiator mounted load bank

Table 6: Generator Potential Sub-Systems

Considerations

- Generators might be tested in the early stages of building construction, possibly before the normal power service is energized. Be sure to walk the project site and raise awareness before generator operation, especially if lights will blink due to open transitions (safety).
- Review IEEE 446 (Orange Book) for reverse power and additional considerations.
- For custom enclosures, ensure the design is coordinated. For example, does the battery charger have a time delay relay to avoid alarming (due to power loss) during open transition transfers?
- If the provided system has reporting capability, commissioning is the perfect time to validate the output. For example, in healthcare, monthly testing is required for emergency generation and transfer switches. If specified, a time saving pre-formatted report is generated after the test is complete.
- If the generator service technician is on-site during commissioning, they may be able to tether the generator controller directly to their laptop for trending the data at a time interval of the Commissioning Agent's choice. After the test, the generator technician then exports a spreadsheet for the test record. This information may be useful to compare against the PQM output.
- Before starting the generator, make sure any automated dampers operate correctly. If the louvers do not open before the generator starts moving air, undesirable outcomes may occur.
- Alternator protection and feeder sizing considerations may apply if circuit breakers are remote.

- Consider load banking activities and have discussions with the utility provider before commissioning to avoid unnecessary demand charge ratcheting, as the total load might be placed on normal power at some point during functional testing.
- To avoid wet-stacking, limit running diesel generation with less than 30% load for extended periods. Likewise, limit the running of diesel generation according to the Tier rating. The Manufacturer warranty may have annual run and maximum loading limitations. These conditions typically restrict non-emergency (loss of utility) operation to 100-200 hours per year.
- Parallel installations should review load sharing. The most common approach for high power factor building applications is resistive-only load banks for on-site testing.
- Consider if the system is separately derived or not (grounding/bonding).
- Commissioning may be a good opportunity to test storm avoidance systems. Be sure to consider Tier rating and EPA allowances as applicable.
- 2017 NEC requires monitoring of the start signal for emergency systems. The start wiring should be isolated from other wiring. Pathways may require protection by a rated system.
- See the Planning section for voltage dip considerations, which may require external control power for the on-site load banks rather than deriving this controller/fan power from the generator.

Uninterruptible Power Supplies (UPSs) and (Central) Emergency Lighting Inverters

See the sections earlier in this chapter for further insights.

Brief Description

UPS equipment provides high availability and power quality assurance. Static (no moving parts) UPS systems are either single-fed or if the static bypass requires a separate feeder for redundancy, possibly dual-fed. Systems are line-interactive (parallel online) or double-conversation (series online). 4W+G output is common for High-Performance Computing (HPC) systems (supercomputers). UPS form factors range from below desk to entire dedicated equipment rooms and potentially separate battery rooms. UPS power factor ranges from 0.75 to unity (1.0). Design elements have enhanced requirements if energy storage is in the form of batteries. Rotary UPSs have a rotating mass to hold the load during normal operation and do not need to recreate the sine wave on the output like a static UPS.

Lighting inverter systems are similar to UPSs, except upon power loss, inverters may have a delay before energizing the load. Since emergency loads have ten seconds to return, this meets code and may be less expensive than an uninterruptible source (NEC 700). Some UPS systems are listed for emergency lighting applications. Instead of a central lighting inverter, the design could choose a UL 924/UL1008 local device and battery back at each emergency light fixture. NFPA 111 applies to supplies 500VA and larger.

Abbreviations

UPS, DRUPS-E/M, EM INV

Types

- Static UPS
  - Transformer-less
  - Transformer-based
- Diesel Rotary (Motor-Generator) UPS
  - Electrically coupled
  - Mechanically coupled

Codes/Standards

In addition to the notes in this chapter's earlier sections: NEC 480, NEC-Chapter 7, NECA 411, IFC, NFPA 1, NFPA 70B-Chapter 28, NFPA 111, NFPA 855, UL 924, UL 1642, UL 1778, UL 1973, UL 9540, UL 9540A, OSHA Part 1926.441, IEEE 493, IEEE 399, IEEE 1679.1, and IEEE 1188.

Key Functional Tests/Verifications

NFPA 111 requirements for Level 1 systems	Step loads
Monitored points	Block loads
Settings confirmations	Emergency Power Off (EPO) and manufacturer safeties
All modes: online (inverter), off-line (static-bypass), maintenance bypass (external or in-line), economy (module management, advanced economy), recharge, restart	Modular power modules: adding and removing while de-energized
Failure schemes: opened battery breaker, loss of rectifier and/or static bypass power (inputs)	Optional: momentary overloads (recommended for factory testing rather than on-site)
Repeat all modes: while sourced from the generator power – it is critical this test is performed	Battery discharge/charge
Repeat all modes: while in parallel, if applicable	Extended burn-in

Table 7: UPS / INV Functional Testing

Measurements

Equipment total power consumption from the local controller, input/output PQMs, and load bank. Consider where the output PQM meter is to be hooked-up in the maintenance bypass panel.	Battery voltages and temperature, for both strings and individual jars – review IEEE 1188 for VRLA batteries
Efficiency	Electrical trending
Time it takes to transfer for comparison against NEC-Chapter 7 requirements, if applicable	Infrared (IR) scanning of terminations and DC link interconnects
Flywheel or rotary vibration	Sound level
Transients during step loads	Transients during block loads

Table 8: UPS / INV Test Measurements

Sub-Systems

Remote EPO	Remote annunciator
Network card	DC ground fault detection
Sharing inductor	Battery monitoring system
Rotary automatic greaser	UPS paralleling equipment
Rotary pony motor	Rotary helium or vacuum system
Isolation transformers	Solenoid Key Release Unit (SKRU) and related interlocks
External or in-line bypass cabinet	Central or distributed static switches
Energy storage (DC Link). Flywheel: magnetic or ball bearings. Batteries (top or side terminal): VRLA, thin-plate VRLA, wet cell, lithium-ion	

Table 9: UPS / INV Potential Sub-Systems

Considerations

- An NEC 702 (optional standby) UPS system cannot serve as UL 924 (90-minute) emergency back-up. A dedicated UL 924/UL1008 option is best to comply with all codes.
- Consider if duplicate kirk-keys may need to be destroyed for the sake of safety.
- Ideally, the DC link voltage has dissipated before the UPS is officially shut-off.
- The maintenance bypass two or three-breaker system and SKRU protection must be reviewed closely. The bypass cabinet is the location that combines the UPS output with a utility source. To perform manual bypass, this operation requires similar electrical characteristics for both sources. The commissioning process must also confirm the kirk-keys have been installed on the appropriate breakers to isolate power safely.
- Parallel installations should review load balancing in all UPS modes. In particular, for distributed static switch designs, the bypass mode requires equal feeder lengths or sharing inductors to maintain equal load sharing.
- Most UPS systems start-up and immediately go on-line (turn on the inverter). Use caution with downstream transformers, as the system should be in maintenance bypass to avoid in-rush on the UPS inverter. Shunt or undervoltage tripping of breakers in the UPS downstream distribution may be prudent to force the UPS to be restarted in a specific sequence.
- Consider if the system is separately derived or not (grounding/bonding).
- Runtime de-rating may be required if battery room temperatures are designed for over 77F (VRLA or flooded cell, IEEE 1188). Especially, consider the lithium-ion battery warranty and confirm with the Vendor before assuming higher room temperatures are allowed.

- The UPS requires "on-generator" signal wiring. Although noted as a product feature, most design specifications do not detail this in the execution (Part 3) portion of the specifications. The wiring is commonly missed and is needed for the following reasons:
  - It allows the user to pre-set a reduced DC link charging rate while on generator; this setting reduces the fuel usage while the system is in an abnormal state. As a percentage of the UPS's frame size, a common normal charging rate is between 10% to 25% and the reduced setting would be lower. This setting may need to be enabled by a service level password. The rectifier feeder and overcurrent protection must also accommodate sizing for load plus charging current.
  - It informs the UPS that the input source is not as stiff (12%+ sub-transient reactance for a back-up generator compared to roughly 5.75% utility transformer impedance), so a wider range of input imperfections without transferring to static bypass is allowed by the controller (stay on-line to protect the load).
  - When utilizing economy modes, if the UPS system becomes powered by a generator source, the industry best practice is for the UPS controls to automatically return on-line (inverter).
- For successful functional testing, these design considerations are crucial:
  - Type of UPS and predicted load percentage while on generator. A size generator/UPS ratio of up to two-to-one may be necessary. For example, a 750kW genset may be required to power a 350kVA UPS; alternatively, a static transformer-less UPS design may be sized very close to the generator rating.
  - To avoid generator leading power factor and shutdown due to reverse power, review if the UPS will be lightly loaded (30% or less) and the input filters need to be cycled off when powered by the generator.
  - Dual feeding UPS systems from separate sources may be accomplished by working closely with the UPS Manufacturer, but this is avoided if possible.
  - Confirm the generator sub-transient reactance is appropriate for the UPS system. 12% or less is desirable but not always achievable.
    - Test the combination in the generator Manufacturer's free consumer sizing software and discuss the possibility of a generator having higher sub-transient reactance with the UPS Manufacturer.
- Operational consideration: rotary UPS designs inherently include a phase shift between the input/output, so the DC link is discharged to complete the internal bypass operation.
- The industry debates whether to do complete battery discharges or not. For VRLA and LIB types, if recharging starts immediately after, this guide suggests the Day 1 battery discharge does not significantly impact the battery's life. Performing this test provides the Owner with how much time they have to respond to input power loss. More than one battery rundown test is not recommended.

<u>Transfer Equipment</u>

See the sections earlier in this chapter for further insights.

Brief Description

Transfer equipment involves packaged or custom distribution that accepts two or more input sources, with or without isolation ties. Bypass versions may be available for packaged equipment (redundant manual switch inside the equipment). These are commonly stand-alone pieces of equipment but may also be custom designed by the EOR/Manufacturer inside switchboard or switchgear. When designing custom equipment, consider UL 1008/UL 1008A listing requirements. Transfers occur manually or automatically to power downstream loads. Except for STSs, all transfer switch technologies have a momentary loss of power to the downstream load upon unexpected loss of primary source power. Service entrance options are also available, which provide a neutral-ground bonding point.

Abbreviations

ATS, MTS, ATO, STS

Types

- Pre-Packaged Controller
  - Automatic Transfer Switch:
    - Automatic sensing and transfer
    - Breaker or contactor-based
  - Non-Automatic Transfer Switch:
    - Electronic transfer via human operator pressing controller buttons
    - Breaker or contactor-based
    - Sometimes referred to as an MTS
  - Static Transfer Switch:
    - Similar to an automatic switch, but with power electronics performing electrical switching (no moving parts) instead of contactor or breaker-based transitions (moving parts).
    - 1/4 cycle transfers (4ms, no-blip)
    - Molded case switches are commonly used for manual bypass
  - Optional: Maintenance bypass feature
- Automatic Throw-Over
  - Custom or pre-packaged controller
- Manual Transfer Switch
  - Human operator, often with kirk key interlocks on circuit breakers
- Construction: 3 or 30 cycle

- Transition Type:
  - Open, closed, overlapping neutral, delayed
  - Delayed transition may be necessary for motor decay
  - Closed transition should be confirmed with the local utility

Codes/Standards

In addition to the notes in this chapter's earlier sections: NEC, UL 1008, UL 1008A, and ANSI.

Key Functional Tests/Verifications

NFPA 111 and NFPA 110 requirements for Level 1 systems	Settings confirmations
Monitored points	Loss of primary source
All modes: manual, automatic, run generator, test, bypass switch, bypass timer, failure modes, etc.	Loss of all sources
All transition options	Extended parallel lock-out (100ms)
Power-on and transfer to the alternate source if the primary source is not available, etc.	

Table 10: Transfer Equipment Functional Testing

Measurements

The time it takes to transfer for comparison against NEC-Chapter 7 requirements

Sub-Systems

Metering	Potentially differential protection systems
Network card	Engine start signals
Inhibit signals	Remote annunciator
Infrared scanning for connections less than 40 cal/cm^2	Multiple incoming sources
Potentially station power and battery charger	Dry-contacts for BMS and UPS monitoring and/or control
Elevator, fire alarm system, or similar interconnections	

Table 11: Transfer Equipment Potential Sub-Systems

Considerations

- STSs are becoming less common since IT rack equipment often uses dual cord power supplies. However, there are many other inherent benefits to this technology that should be considered. For example, fault tolerance and overload protection.
- The primary and alternate sources, except on custom equipment, are usually not programmable. The source-one feeder must land on the source-one lugs, and the desired source-two must land on source-two lugs. For example, an automatic switch with a packaged controller eventually (due to max time delay) returns to source-one since it is the primary input. Design Engineers should make the source 1/2 distinction clear on the drawings.
- Note front/rear/side access requirements based on where the equipment panels need to be removed for terminations and to have access for IR scans.
- UL 1008, as of the seventh edition, no longer allows similar breaker TCCs to be utilized for compliance. Every individual breaker and trip unit must be a tested-combination to comply with the listing (associated with the kAIC). The medium-voltage equivalent UL 1008A may have similar restrictions. 2017 NEC-Chapter 1 includes requirements to list kAIC ratings on the face of service equipment.
- The system may initiate a start signal when racking the bypass isolation back-in if the transfer equipment loses power during this activity (no back-up source).
- Review the design to see if there are other ATSs or STSs integral to the mechanical equipment, such as in-row cooling for computer room units that accept two sources of power for local redundancy. These may benefit from testing during the commissioning process.
- Consider if room separations are required, as defined by NFPA 110.
- Consider if neutral connections should be broken for source transfers – this will depend on separately derived or not (grounding/bonding).
- For custom transfer equipment, consider the number of no-load vs with-load switching operations allowed before the Manufacturer requires maintenance.
- Confirm kirk-key systems operate exactly as their posted sequence describes and that the correct breakers are locked to isolate safely.

Mike Starr, PE

Annex A – Document Summary

Commissioning may be looked at from multiple perspectives. For this Annex, the process is summarized with respect to the physical paperwork that might be needed. The documents listed apply to a standard design-bid-build process. An IPD project may combine documents or have shared contributors. The kinds of documents and scope/scale need to align with the project. The lists are not all-inclusive.

This focused listing overviews the majority of the documents associated with the *electrical* commissioning process. Depending on the project, the documents may have different names. For example, the Test Equipment Plan may also be referred to as a Load Bank Plan. Given the number of documents, quality control and revision tracking processes are needed. The document owners should also consider when the deliverables need to be available for team review in order to maintain schedule.

The lists are based on which party typically creates the document. For instance, Level 3 IVCs are usually written by the Commissioning Agent but completed by the Installing Contractor. Some items listed are grouped terms. For example, VDC includes multiple deliverables not listed, such as 3D models, 2D CAD files, clash meeting resolution logs, etc.

Some documents are not identified. For example, Project Teams that host meetings produce meeting minutes. Projects likely also have multiple contributors for lessons-learned, value engineering, responsibility matrices, contract/legal documents, field/site observation reports, etc. See the Commissioning Process chapter for further insights.

This annex does not attempt to distinguish between standard and enhanced project services. Please coordinate with project providers and partners to confirm the included scope for your application. Likewise, Owner and Design Teams must coordinate the desired acceptance testing services.

Owner

Facility Standards	Standard Operating Procedure (SOP)
Owner Project Requirements (OPR)	Method of Procedure (MOP)
Request for Proposals (RFP)	Current Facilities Requirement (CFR)
Design Review Comments	Utility Management Plan (UMP)
Historical Utility Usage Data	Site Labeling Guide
Request for Quotes (RFQ)	

Table 12: Owner Document Summary

Design Team

Design Proposal	Construction Documentation QA/QC
Codes and Standards Review	Energy Compliance Documentation
Basis of Design (BOD)	Record Drawings
Design Plans	Commissioning Review Comments
Design Specifications	Final Punchlist
Design Opinion of Probable Cost	Submittal Review Comments
Design Sequence of Operation (SOO)	Request for Information (RFI) Responses
Design Electronic Files (Models, Studies, etc.)	

Table 13: Design Team Document Summary

Equipment Manufacturers

Factory Safety Plan	Shop Drawings
Factory Acceptance Testing	Product Data
Factory Witness Test (FWT) Plan	Manufacturer Sequence of Operation (SOO)

Table 14: Equipment Manufacturer Document Summary

Installing Contractors

Primary Construction Schedule	Contractor Quality Plan
Submittals / Shop Drawings	Owner Training Plan
Power System Study	Acceptance Testing Data
Requests for Information (RFIs)	Infrared (IR) Scan Report
Torque Logs	Operations & Maintenance Manual (O&M)
Site Safety Plan	As-Built Drawings
Construction Virtual Design and Construction (VDC)	Lift Plans

Table 15: Installing Contractor Document Summary

Commissioning Agent

Commissioning Plan	Installation Verification Checklists (IVCs)
Commissioning Schedule	Functional Performance Tests (FPTs)
Commissioning Specifications	Integrated System Test (IST)
Test Equipment Plan	System Manuals
Commissioning-Level Design Review Comments	Power Quality Meter (PQM) Report
Commissioning-Level Submittal Review Comments	Final Commissioning Report
Requests for Information (RFIs)	Seasonal Testing Report
Issue / Discrepancy Log	

Table 16: Commissioning Agent Document Summary

Annex B – Acronyms and Abbreviations

A	Ampacity or Amperage
AC	Alternating Current
ADA	Americans with Disabilities Act
AHJ	Authority Having Jurisdiction
AIA	American Institute of Architects
ANSI	American National Standards Institute
APF	Apparent Power Factor (Distortion Power Factor)
APFC	Apparent Power Factor Correction
ARMS	Amps RMS
ASHRAE	American Society of Heating, Refrigeration and Air-Conditioning Engineers
ASTM	American Society for Testing and Materials
ATO	Automatic Throw Over
ATS	Automatic Transfer Switch
ATS	(NETA) Acceptance Testing Standard
AWG	American Wire Gauge
BCA	Building Commissioning Association
BFB	UPS Back-Feed Breaker
BIL	Basic Insulation Level
BIM	Building Information Modeling
BMS	Building Management System
BOD	Basis of Design
C	Capacitance
C	Celsius
CA	Construction Administration
CAD	Computer Aided Design
CAFM	Computerized Asset and Facility Management
cal/cm^2	Calories per Centimeter Squared
CAT	Category
CBEMA	Computer Business Equipment Manufacturers Association
CCW	Counter-Clockwise (Rotation)
CD	Construction Documents
CFD	Computational Fluid Dynamics
CFR	Current Facilities Requirement
CFR	Code of Federal Regulations
CHP	Combined Heat and Power
CIR	Corrective Issues Report
CM	Construction Manager

CMMS	Computerized Maintenance Management Systems
CSI	Construction Specification Institute
CRAC	Computer Room Air Conditioner
CRAH	Computer Room Air Handling Unit
CT	Instrument Current Transformer
CW	Clockwise (Rotation)
Cx	Commissioning
D	Distortion (VAR)
DC	Direct Current
DCC	Data Center Continuous Rating
DCiM	Data Center Infrastructure Management
DCS	Distributed Control Systems
DD	Design Development
DLO	Diesel Locomotive Cable
DOE	Department of Energy
DPF	Distortion Power Factor (Apparent Power Factor)
DRUPS	Diesel Rotary Uninterruptible Power Supply
DSCADA	Distributed Supervisory Control and Data Acquisition
EM	Emergency
EOR	Engineer of Record
EPA	Environmental Protection Agency
EPC	Engineer Procure Construct
EPO	Emergency Power Off
f	Frequency
F	Fahrenheit
FAT	Factory Acceptance Testing
FEMA	Federal Emergency Management Agency
FGI	Facility Guidelines Institute
FM	Factory Mutual
FPP	Functional Performance Procedures
FPT	Functional Performance Testing
FWT	Factory Witness Testing
GC	General Contractor
GFCI	Ground Fault Circuit Interrupter
GSA	US General Services Administration
HMI	Human Machine Interface
HPC	High-Performance Computing
Hz	Hertz
IAT	Integrator Acceptance Testing
IBC	International Building Code

ICEA	The Insulated Cable Engineers Association
ID	Identifier
IEC	The International Electrotechnical Commission
IEEE	The Institute of Electrical and Electronics Engineers Standards Association
IESNA	Illumination Engineering Society of North America
IFC	International Fire Code
INV	Inverter
IPD	Integrated Project Delivery
IR	Infrared Scan
ISO	International Organization for Standardization
IST	Integrated System Test
IT	Information Technology
ITI	Information Technology Industry Council
IVC	Installation Verification Checklist
IWT	Integrator Witness Test
k	Kilo (thousand)
kAIC	Kilo Ampere Interrupting Capacity
kCMIL	Kilo Circular Mil
kV	Kilo Volt
kVA	Kilo Volt-Amps
kW	Kilo Watts
kWH	Kilo Watt-Hour
L	Inductance
LED	Light-Emitting Diode
LEED	Leadership in Energy and Environmental Design
LIB	Lithium Ion Battery
LOTO	Lock Out Tag Out
LV	Low Voltage
MBB	Maintenance Bypass Breaker
MEP	Mechanical Electrical Plumbing
MIB	Maintenance Isolation Breaker
MLO	Main Lug Only
MOB	UPS Module Output Breaker
MOP	Method of Procedure
ms	Millisecond
MTM	Main-Tie-Main
MTS	(NETA) Maintenance Testing Standard
MTS	Manual Transfer Switch
MV	Medium Voltage

MW	Megawatt
NEC	National Electrical Code
NECA	National Electrical Contractors Association
NEMA	National Electrical Manufacturer's Association
NESC	National Electrical Safety Code
NETA	National Electrical Testing Association
NFPA	National Fire Protection Agency
NIC	Network Interface Card
NICET	National Institute for Certification in Engineering Technologies
NIST	National Institute of Standards and Technology
O&M	Operations & Maintenance
OFCI	Owner Furnished Contractor Installed
OPR	Owner Project Requirements
OSHA	Occupational Safety and Health Administration
OSHPD	California's Office Statewide Health Planning and Development
P	Real Power
PCS	Process Control System
PDF	Portable Document Format
PDU	Power Distribution Unit
PE	Professional Engineer
PF	Power Factor
PLC	Programmable Logic Controller
PM	Preventative Maintenance
PM	Project Manager
PMS	Power Monitoring System
PPE	Personal Protective Equipment
PQM	Power Quality Meter
PSU	Power Supply Unit
PUE	Power Usage Effectiveness
Q	Reactive Power (VAR, Leading or Lagging)
QA/QC	Quality Assurance/Quality Control
R	Resistance
RDF	Relay Database File
RFAT	Remote Factor Acceptance Testing
RFI	Request for Information
RFP	Request for Proposal
RFQ	Request for Quote
RLC	Resistor-Inductor-Capacitor

RMS	Root-Mean-Squared
ROM	Rough Order of Magnitude
RPM	Revolutions Per Minute
S	Apparent Power
SAT	Site Acceptance Testing
SCADA	Supervisory Control and Data Acquisition
SD	Schematic Design
SKRU	Solenoid Key Release Unit
SOO	Sequence of Operation
SOP	Standard Operating Procedure
SMPS	Switched Mode Power Supply
SRC	System Readiness Checklist
STS	Static Transfer Switch
SWP	Standard Work Practice
TCC	Time Current Curves
THD	Total Harmonic Distortion
TIA	The Telecommunications Industry Association
TRV	Transient Recovery Voltage
UIB	UPS Input Breaker
UL	Underwriters Laboratory
UMP	Utility Management Plan
UPS	Uninterruptible Power Supply
US	United States of America
USB	Universal Serial Bus
USS	Primary or Secondary Unit Sub Station
V	Voltage / Volts
VDC	Virtual Design and Construction
VDC	Voltage, Direct Current
VE	Value Engineering / Value Enhancement
VFD	Variable Frequency Drive
VLF	Very Low Frequency
VRLA	Valve Regulated, Lead Acid
VRMS	Volts RMS
VTP	Verification Test Procedures
W	Watts or Wattage
WELL	Well Building Standard
WP	Weatherproof
X	Reactance
Z	Impedance
ZSI	Zone Selective Interlocking

Made in the USA
Middletown, DE
22 January 2024

48198992R00080